Future War

An Assessment of Aerospace Campaigns in 2010

JEFFERY R. BARNETT

Air University Press
Maxwell Air Force Base, Alabama

January 1996

Library of Congress Cataloging-in-Publication Data

Barnett, Jeffery R., 1950–
    Future war : an assessment of aerospace campaigns in 2010 / Jeffery R. Barnett.
      p. cm.
    Includes bibliographical references and index.
    1. Air warfare—Forecasting. 2. Twenty-first century—Forecasts. I. Title.
    UG632.B37   1996
    358.4′009′0501—dc20   95-54022
                                                   CIP

Disclaimer

This project
would have been impossible
without the support and advice of my wife,
former Captain Katherine Hoyland Barnett, USAF.

Contents

Illustrations

Photographs

Foreword

Jeff Barnett, one of a growing number of military professionals who recognize that the United States has entered a new revolution in military affairs, thinks carefully about what this revolution means for the US military, where it will lead us, and what must be done with it to give the nation a new level of military power in the twenty-first century. He has some well-articulated insights.

Colonel Barnett's prognostication: It will take hard work to ride the revolution to its promise, and hard thinking, now, to plot the proper course. While the revolution may already be underway, the United States, in his view, will not be its sole beneficiary. Nor will the United States monopolize the technical and doctrinal engines that drive this revolution forward; others will also seize its driving technologies. As a result, the United States must be prepared to face both peer and niche competitors in the future—the former with military capabilities roughly equal to those of the United States, the latter with capabilities perhaps sufficient to deter US intervention even when important US interests are at stake.

The author therefore focuses on one of the central issues in the emerging US debate on national security; namely, to what extent can the United States use its lead in the new revolution in military affairs to deter the growth of peer and niche competitors, or defeat them if they decide to compete directly with military force. His answers are straightforward: We must develop superior capabilities in information war, joint operations, and parallel war. He lays out what this means in specific technological and doctrinal terms.

It is here, I think, that the author makes his greatest contribution. His assessment is free from the kind of hyperbole that sometimes creeps into discussions of the changes under way in the US military, and he lays out in direct and balanced form the promises of information war theory and the leverage it gives to the concept of parallel war. But he also indicates the kinds of changes that will be necessary if we are to achieve this in the face of determined opposition by peer and niche competitors sometime early in the next century. The changes are pretty drastic, as suggested by the following sample:

- radical alterations to current air defense operations
- CONUS-positioned command
- low signature becomes key to successful military operations
- the potential of stealthy cruise missiles to dominate warfare
- the probable end of large invasion forces.

This is a book about the future and about aerospace war in the future. It is a book all warriors should read.

WILLIAM A. OWENS
Admiral, USN
Vice Chairman, Joint Chiefs
of Staff

About
the Author

Colonel Jeffery R. Barnett

Colonel Jeffery R. Barnett, USAF, is the Military Assistant to the Director, Net Assessment, Office of the Secretary of Defense.

Colonel Barnett is a senior navigator with 2,000 flying hours in the C-130 and the C-9. He flew 40 combat and combat-support missions in Southeast Asia. He also flew as a member of the personal aircrew for the Supreme Allied Commander, Europe. His staff assignments include Headquarters United States Air Force and Headquarters Pacific Air Forces. During the 1991 Gulf War, he was attached to the CHECKMATE planning cell.

Colonel Barnett earned a bachelor of arts degree in history from the College of the Holy Cross and holds a master of science degree in international relations. He is a graduate of the Air War College, the Air Command and Staff College, Squadron Officer School (Distinguished Graduate), and the Army-Air Force Air/Ground Operations School (Distinguished Graduate).

Colonel Barnett has published widely. His recent publications include: "The Shadows on Veterans' Souls" *(Houston Post,* November 11, 1994); "War's Worst Risk Is to the Soul" *(Los Angeles Times,* June 6, 1994); "Nonstate War" *(Marine Corps Gazette,* May 1994); "Exclusion as National Security Policy" *(Parameters,* Spring 1994); and "We Have More to Offer" *(Proceedings,* April 1993).

Preface

The purpose of this book is to outline the aerospace aspects of future war. Because future war is an exceptionally broad subject, three caveats are in order.

1. This book outlines only future *state versus state* warfare. Its theories are applicable only to future wars between sovereign states and alliances of sovereign states. States have organized militaries, infrastructures, production bases, capitals, and populations. These components enable unique capabilities and vulnerabilities—which dictate the scope and character of war. Because states alone have these attributes, theories of state versus state war are unique.

The book is *not* intended to provide a template for wars with nonstates such as future versions of Somali clans, Bosnian Serbs, or Vietcong. Nonstate warfare is certainly important; its future deserves serious treatment. However, because nonstates differ fundamentally from states, an examination of future nonstate warfare requires a wholly separate treatment. Nonstates, by definition, exist without infrastructures, production bases, and capitals. Nonstates usually have neither organized militaries nor any responsibility for populations. In essence, nonstates have completely different makeups relative to states. Because of these gross differences, nonstates require their own theories of war.[1] It is impossible to reconcile both state and nonstate conflict into one theory. Future aerospace operations in wars with nonstates must remain for others to address. This particular book views future aerospace operations through only one prism, that of state versus state conflict.

2. This book reviews only the *aerospace aspects* of future war. This limited focus is not meant to slight land and naval campaigns—they will remain crucial to future war, forming fundamental components of joint campaigns. However, land and naval campaigns are extensive in their own rights. Their projections must remain for others to outline. This book will

project land and naval warfare only to the extent that (1) they are affected by aerospace operations and (2) armies and navies are supported by, or operated in support of, aerospace campaigns.

3. This book begins and ends in the future. It is *not* a linear extrapolation of current trends; it is not meant to summate the cumulative effects of current policies and programs. Rather, this book outlines how future warfare could be conducted, given the expected advances in technology. Whether or not national decision makers choose to equip, organize, and train aerospace forces in accordance with this vision of the future is a separate question—one that is beyond this book's scope.[2]

Notes

1. Col Jeffery R. Barnett, "Nonstate War," *Marine Corps Gazette*, May 1994, 84–89.

2. This book also does not recommend specific treaty changes (e.g., intermediate-range nuclear force treaty, antiballistic missile treaty).

Acknowledgments

Credit for many of the ideas in this book rests with others. Far and away the most creative and insightful thinkers on future warfare are Andrew W. Marshall, Director, Net Assessment (Office of the Secretary of Defense) and Col John A. Warden III, USAF, Retired, former Commandant, Air Command and Staff College. They inspired key concepts in this book. Other significant contributors include Col Tom Smith, USA; Capt Jim FitzSimonds, USN; Col Chuck Miller, USAF; Cmdr Jan van Tol, USN; Dr Tom Welch; and Col Dick Szafranski, USAF. The insights and encouragement of Dr Andrew Krepinevich; Capt John Hanley, USNR; Dr Michael Vickers; David Ochmanek; Fred Frostic; Dr Chris Bowie; Capt Dick O'Neill, USN; Maj Scott Dorff, USAF; Mark Herman; Col Don Selvage, USMC; Col John Roe, USAF; Dr Fred Giessler; Dr Bruce Bennett; Dr Lou Marquet; Maj Terry Fox, USAF; Lt Gen Dale Vesser, USA, Retired; Col Dave Deptula, USAF; Dr Mike Brown; Dr Chuck Bartlett; Dr Barry Watts; Maj Joe Squatrito, USAF; Capt Jim Hollenbach, USN; and Col Jamie Gough, USAF, were invaluable. My editor at Air University Press, Mr Preston Bryant, was professional in every respect.

Introduction

As we look to the future of war we must face one absolute certainty: any projection will prove faulty. Despite our best intellectual efforts, the future will remain unknowable. Between now and the focus of this study, the year 2010, humankind's innumerable decisions will interact to form a future far beyond our powers to predict. It would reveal the greatest hubris to claim absolute insights on such a dynamic, multidependent future.

This limitation is especially true concerning war. Any projection of future war must contain implicit assumptions of time, enemy, location and purpose. We must know *when* the war occurs in order to project what kinds of technologies might be available. We must know the *purpose* of the war in order to forecast what level of effort the Nation will dedicate. We must know the *enemy* in order to build the most appropriate strategic campaign, addressing both offensive and defensive centers of gravity. Finally, we must know the *location* of the war in order to define the types and numbers of operational targets. All four factors will interact to define the nature and conduct of any future war. However, because *none* of the factors are knowable in advance, any vision of future war will be severely limited.

Accepting these limitations, this book will center on two themes. First, it will highlight where fundamental changes in military operations have already occurred. Militaries, by their nature, are hesitant to embrace unproven theories. As a result, they are usually slow to recognize new possibilities in operational art. Nonetheless, fundamental change in the conduct of war is a constant throughout history. To stay ahead of this change military planners must constantly reevaluate their concepts of war. This book is designed to help military professionals recognize new opportunities mandated by changes that have already occurred in the technological and political environments.

The second theme in this book is the impact of foreseeable technological advances on military operations. Technology is not standing still; if anything, technological advances are accelerating. Significant advances in technology are a valid planning assumption over the next 15 years. Advances in microprocessing and all its supporting technologies will drive new possibilities on future battlefields. While the exact specifications of these technologies are beyond our capacity to define in advance, we can assume technological change of at least the pace experienced over the past 10 to 20 years. Technological change of this magnitude will mandate commensurate changes in military operations. This book will explore some of the more significant impacts of probable technologies on the future battlefield.

Each argument exploring these twin themes is explained in detail in the succeeding chapters. Understanding the need for some readers to have an overview of key points in advance, the following paragraphs will introduce the key arguments. They are not intended to convince skeptical professionals of the validity of each proposal. That is the purpose of the detail in the succeeding chapters.

Despite current optimism, a *peer competitor* to the United States will eventually arise. Only the timing is unknown. Due to the time needed for tensions to increase and rearmament to begin, and based on historic intervals of fundamental change in security and technology, the earliest edge of this window is approximately 15 years (2010).

A peer competitor is defined as a state (or alliance) capable of fielding multiple types and large numbers of both emerging and present weapons, then developing an innovative concept of operations (CONOPS) to realize the full potential of this mix. In most ways, a peer's military capabilities will roughly equal those of the United States. The peer's goal will be to control a vital interest of the United States, on either a global or regional basis, then defeat the US military response. Historic examples of peer competitors include the USSR, Nazi Germany, and Imperial Japan.

A *niche competitor* is defined as a state (or alliance) that combines limited numbers of emerging weapons with a robust inventory of current weapons, then develops an innovative concept of operations to best employ this mix. The niche's overall military forces will be inferior to those of the United States. Its goal will be to effectively challenge US interests in its region by making the US military response sufficiently costly to either deter initial involvement or dissuade further involvement on the part of the US.

A niche competitor could arise at any time over the next 10–20 years. It could access (1) civilian space networks for surveillance and communications, (2) international arms markets for low-observable missiles with precision guidance, and (3) computer and communications professionals for information war. Although these technologies are only emerging today, they'll likely become widely available over the next 10–20 years. By 2005–2015, many countries could obtain these technologies and integrate them with the rest of their military. Examples of possible niche competitors include Iraq and North Korea.

In general, the military capabilities of future peer and niche competitors will differ significantly in both quality and quantity. While both will incorporate emerging advances in the key technologies of information, command and control (C^2), penetration, and precision, they will do so in markedly different ways.

For example, a peer competitor will field surveillance systems that are dedicated to military applications. They will respond directly to the peer's tasking. Taskings will cycle in near real time (NRT). A niche competitor, on the other hand, will likely depend more on commercial information systems (e.g., commercial communications satellites). These systems will be less responsive, especially when controlled by a third party. A niche will experience greater time delays between data collection and dissemination to weapon systems. Peers will pose multiple types of challenges to defense systems, while niches will confront defenses with only a few types of penetrating systems. The niche will also depend more on off-board guidance (such as the Global Positioning System) while a peer will use autonomous guidance systems (such as inertial guidance).

A peer will have the wherewithal to avoid being defeated by a single, crushing blow by the United States. It will have sufficient depth and wealth to preclude being overwhelmed by massive numbers over a small area. It will also deter strategic attacks with a robust nuclear retaliatory capability. Conversely, a niche competitor must contemplate war without these advantages. Table 1 summarizes the major differences between peer and niche competitors.

Table1

Major Differences Between Peer and Niche Competitors

	Peer Competitor	Niche Competitor
Information	Indigenous, Dedicated	Third Country, Commercial
c^2	NRT, Redundant, Automated	Delayed, Nodal, Hierarchical
Penetration	Multisystem	Single System
Precision	Autonomous Guidance (e.g., terminal)	External Guidance (e.g., GPS)
Weapons of Mass Destruction	Hundreds. Can Reach USA.	<10, Theater Reach
Size	Large, Strategic Depth	Small, Little Depth

Any forecast of future aerospace war must reflect the current *revolution in military affairs* (RMA). Historically, RMAs occur only when militaries fundamentally change both their concepts of operations (CONOPS) and their organizational structures to best employ radically new technologies. In other words, revolutions in military affairs are underwritten by new technologies but realized through new operational and organizational concepts.

Technological advances in four general areas are underwriting a new aerospace approach to future war:

1. Information
2. Command and control
3. Penetration
4. Precision

By 2010, well into the information age, aerospace planners will have an incredible amount of information about the target state. They'll never know everything, but they will detect orders of magnitude more about the enemy than in past wars. With this information, commanders will orchestrate operations with unprecedented fidelity and speed. Commanders will take advantage of revolutionary advances in information transfer, storage, recognition, and filtering to direct highly efficient, near-real-time attacks. Responding in this direction, aerospace attackers will take advantage of advances in stealth, hypersonic, and/or electronic warfare technologies to greatly increase the chances of penetration. While defenses will certainly defeat some attackers, others will get through at rates higher than previously experienced. Finally, once over the target area, aerospace platforms will deliver brilliant munitions. Deliveries will be highly accurate. Target locations will be measured within feet. Working together, advances in these four areas of aerospace technology will underwrite a revolution in military affairs.

The new operational concept to realize the potential of these underwriting technologies is *parallel war.* In parallel war, aerospace forces simultaneously attack enemy centers of gravity across all levels of war (strategic, operational, and tactical)—at rates faster than the enemy can repair and adapt. The overall goal of parallel war is paralysis of the enemy through shock (as opposed to gradual attrition). For this reason, leadership is the highest priority target. Once paralyzed, the enemy will be unable to orchestrate either a damaging offense or an effective defense.

Parallel war requires large numbers of highly precise weapons directed against vital targets. While this concept has long been envisioned by strategists in theory, advances in technology are currently enabling its prosecution in reality. Aerospace forces will soon be able to engage hundreds of targets within the first hour of a conflict. They will deliver thousands of precision munitions within each 24-hour period. Enabled by advance information systems, these weapons will strike vital enemy targets. The sum of these capabilities drives more than an increase in military efficiency. As explained in chapter 1, these capabilities drive a new concept—parallel war.

New organizational concepts are needed to support these new technologies and concepts of war. The greatest need is for new approaches to *command and control.* This is the number one issue facing today's aerospace planners. The information age is rapidly increasing the amount, speed, and fidelity of data gathered and distributed to war fighters. Exponential advances are on the horizon. However, the basic command hierarchy for US forces has remained roughly the same. While commands move through the system much faster than before, the basic aerospace C^2 system is unchanged. From a historical perspective, this overlapping of new technologies on top of old hierarchies is a signpost for "old think." A more automated and flat structure, notwithstanding its own vulnerabilities, offers the greatest potential for near-real-time, deconflicted, multiservice, multitheater operations.

One major change in aerospace C^2 is needed immediately. The *Joint Force Air Component Commande r* (JFACC) for theater war should remain in CONUS. Basing the JFACC in CONUS would avoid creating a fixed, in-range, high-value target for the enemy. It also would allow immediate planning/tasking of the aerospace campaign. A CONUS-based JFACC would have well-exercised connectivity with combat units (e.g., through fiber-optic cable connections with CONUS-based stealth bomber wings). Target planners would have immediate access to all-source intelligence. All data relayed by satellite (including data from national systems) would down-link to this JFACC which would fuse the data, filter extraneous elements, and distribute distilled information. A CONUS-based JFACC could also take advantage of CONUS databases and expertise; JFACC computers could be hardwired to a secure information net. After running computer simulations to determine the best tactical options, JFACC would issue the air tasking order (ATO). This centralized ATO could direct all air assets, whether based in-theater, in CONUS, or in adjacent theaters. Finally, a single, CONUS-based JFACC would husband the limited number of aerospace strategists and standardize CONOPS, regardless of theater.

Future US theater commanders will call upon a *Joint Force Information Component Commande r* (JFICC) to (1) collect information on enemy capabilities, deployments, and intentions; (2) fuse data collected from all sources and distribute timely information to users; (3) flow friendly information efficiently, even in the face of enemy attacks and competing friendly requirements; (4) degrade enemy information networks; and (5) defend friendly information networks against enemy intrusion. Aerospace forces should expect heavy taskings in support of the JFICC.

Future aerospace operations will require increasingly *centralized execution*. The increasing range of defense weapons and the decreasing range at which stealthy targets will be identified will stress aerospace defenses. Given decentralized C^2, several aircraft—or several batteries—might fire simultaneously at the same target. Or one might shoot while another makes a counterproductive maneuver. Or no one might shoot, everyone thinking someone else has the lead. A centralized execution system can deconflict these factors.

A second factor requiring centralized execution is the multitheater nature of future offensive operations. Routine strike packages may require intelligence and communications from space, bombers from CONUS, escorts from carriers, tankers from a neighboring theater, and unmanned aerial vehicles (UAV) from frontline ground forces. All will operate long-range; all will be interdependent; all will probably receive last-minute changes to their orders. Decentralized execution, effective in past wars, won't answer this challenge.

Because technology will allow near-real-time information and C^2, commanders at all levels will try to move towards *snap decisions* in near real time. This tendency will open interesting opportunities for operational art. Commanders will exploit this tendency. They will make concerted efforts to drive their opponent's snap decisions toward the poor end of the spectrum, usually by presenting false indications of intent or reality. The ultimate goal will be to either slow down the opponent's decision loop or force the opponent to continue making bad decisions in near real time.

Any enemy will have access to these same technologies, of course, and will exploit them to varying degrees. Aerospace planners must prepare for innovative enemies possessing advanced information and C^2 systems, stealthy aircraft and missiles, and precision munitions. Future enemies will also employ the full gamut of existing weapon technologies, including nuclear weapons.

One potentially dominant technology is the stealthy *cruise missile*. Its low signature, independence from fixed launch facilities, and ability to rapidly change routing make its long-range detection very uncertain. Cruise missiles also have the capability to launch from any medium (sea, air, land) and to navigate/target with single-meter accuracy in all weather. They can attack from any direction at any time of day. In addition, prospects for "cheap" stealthy cruise missiles are high. Future militaries will buy/produce thousands of low-observable cruise missiles. Thus, aerospace defenders will have to deal with massive numbers of precise attackers and very little warning.

Such an attack would challenge the number one mission of aerospace forces: to establish aerospace superiority. Thou-sands of stealthy cruise missiles would likely render current aerospace defense CONOPS obsolete. Current aerospace defense CONOPS assume exactly opposite conditions: limited numbers of expensive, high-signature attackers (e.g., Su-24s and Scuds), visible from launch to engagement, with an exposed support infrastructure. Stealthy cruise missiles invalidate those assumptions.

Defensive counters to stealth terminology may be analogous to antitank or antiaircraft developments before World War II. Although huge advances in defense against tanks and aircraft were made before World War II, both tanks and aircraft had decisive effects on that war. Although neither was invulnerable, both dictated a new environment for warfare.

In a similar way, stealth technology will dictate a new warfare environment. It may be possible to degrade the cruise missile's effectiveness by targeting its navigation and terminal area guidance. Air defense units should exploit this potential in addition to improving physical interception capabilities.

One key to successful operations in the emerging environ-ment will be *signature reduction*. All fixed forces with large signatures will eventually be detected and hit. Stealthy cruise missiles and bombers, properly supported by information and precision technologies, will make high-signature, immobile forces extraordinarily vulnerable.

Since air bases have high signatures, aerospace forces should base outside the range of enemy stealth systems. Such basing will be possible only if the aerospace inventory emphasizes long-range operations. Under this condition, aircraft would need long legs for flights from rear area bases to enemy targets. Aircrew ratios would have to support long sortie durations. Aircraft and munitions inventories would have to be sufficient to deliver effective, sustained firepower from distant bases. With minimal support, deployment kits must support extended operations from distant bases. Given these attributes, aerospace forces could effectively operate outside the range of enemy stealth systems.

In the future wartime environment, *large invasion forces* will be highly vulnerable. Their vulnerability will be greatest during the initial stages of an invasion when forces must mass to overcome indigenous defenses. Such wartime standbys as truck convoys, tank columns, ammunition ships that require days to unload, airlift aircraft needing hours to off-load and r efuel, large air bases with "tent cities," and air refueling aircraft parked nose-to-tail may not be possible in the emergi ng environment.

If the United States chooses to oppose an invasion of an ally, it must do so during the initial stages of the attack. Failure to immediately engage the enemy could prove disastrous. If enemy forces gain control of their objective, the US would have to mass forces to expel them.

As US forces mass to capture the lost territory, their logistics, convoys, and buildup areas would come under heavy attack. While the United States will probably be on the strat egic defensive in a future war, it must take the operational offe nsive. Despite any advantages that may accrue to the defense by new technologies, the US can maintain the initiative only through the offensive. In this future environment, US airlift operations must assume significant enemy surveillance.

Space will undoubtedly be a strategic center of gravity in any future war. Both sides will want space control. Whichever side can exploit space for communications, collection, and positioning—while denying similar capabilities to its enemy—will gain a decisive advantage. For this reason, both sides will attack satellites. Attacks on homeland infrastructure for space operations (e.g., launch pads; command, control, communications, computers, and intelligence [C^4I]) may be restricted due to the threat of nuclear retaliation. However, if satellite control depends on a small number of ground stations, those stations will likely come under attack. The United States may operate under certain disadvantages in any future space war. For example, the enemy may

1. shoot first (the National Command Authorities will probably deny first strikes by US forces against enemy satellites);

2. have a power advantage due to a greater willingness to use nuclear power plants in space;

3. have more state-of-the-art satellites due to a faster acquisition cycle;

4. aggressively weaponize space despite US and world opinion; and

5. prosecute an attrition strategy by emphasizing quantity over quality.

To mitigate these disadvantages, planners should integrate satellites and UAVs for communication and surveillance. They should plan for redundancy between types of platforms, overlapping coverage among types of sensors, and connectivity between all platforms with a common C^4I architecture. When combined with innovative C^2, this integration will provide dominant battlefield awareness in a highly competitive environment.

High-signature surveillance platforms will not thrive in the emerging environment. Satellites, especially those in low earth orbit, will be very vulnerable during a future war. As a backup and to heighten crisis stability, the US should prepare a high altitude long endurance (HALE) UAV architecture for com-munication, navigation, and surveillance.

Manned atmospheric platforms, such as AWACS and J-STARS, will have heavy taskings in a war with a niche competitor. In a war with a peer competitor, however, these platforms will prove too vulnerable for operations near enemy forces.

The side that can retrain its entire force to execute the most modern CONOPS will have a decisive advantage in future war. *Computer simulations* may enable this decisive advantage.

Chapter 1

Overarching Concepts

(T)he microchip is transforming how we use air power o n
the battlefield. We now have the ability to acquire an d
communicate huge volumes of information in real time, and
we have the computing power to rapidly analyze those
data. And we have the control systems that allow that
analysis to be passed simultaneously to the users. This is
transforming our battlefield situational awareness and our
battle planning.

—Secretary of Defense William Perry*

As we look to the future, the growth of information technologies seems limitless. Hardly a day goes by without a breakthrough of some kind in information-related technologies. For this reason, it is likely both the US and an enemy will have information-based systems far more advanced than those currently available. Both the US and its enemy could have

- global satellit e networks with voice, data, and imagin g capabilities 50 times greater than today (based on advances i n data compression, processing, frequency management, miniaturization, and sensors). Although the military will control a limited number, commercial interests will own most platforms .
- autonomous weapons, enabled by artificial intelligence, automatic target recognition algorithms, and multispectral miniature sensors.
- sophisticated computer viruses—and equally sophisticated encryption protocols.
- data fusion at rates 10^4 times faster and more accurate than today, based on advances in processing and software. [1]
- data storage capabilities at 10^3 times greater than today (due to miniaturization).

---

*Remarks to the conference of the American Institute of Aeronautics and Astronautics, the Aerospace Industries Association, and the Aviation Week Group, Arlington, Va., 4 May 1995.

Information War

Because of these and other advances, an information campaign will be integral to any future conflict. Simply stated, information will dominate future war. Wars will be won by the side that enjoys and can exploit: cheap information while making information expensive for its opponent; accurate information within its own organization while providing or inserting inaccurate data in its opponent's system;[2] near-real-time information while delaying its opponent's information loop; massive amounts of data while restricting data available to its opponent; and pertinent information while filtering out unnecessary data.

The US does not have a monopoly on this insight. The impact of information technologies on war is well understood abroad. According to one Chinese defense intellectual: "(I)n hi-tech warfare, tactical effectiveness no longer depends on the size of forces or the extent of firepower and motorized forces. But more on the control systems over the war theater and efficiency in utilizing information from the theater. Superiority in numbers and strength no longer plays a decisive role."[3]

Military theorists in Russia hold a similar view. Maj Gen Vladimir I. Slipchenko, (Retired) the leading Russian military theoretician: "The impending sixth generation of warfare, with its centerpiece of superior data-processing to support precision smart weaponry, will radically change military capabilities and, once again, radically change the character of warfare."[4]

Military professionals should feel comfortable envisioning campaigns focused on information. Although an information focus brings new targets and weapons to war, it nevertheless mirrors traditional military concepts in at least six general ways.

1. As with all forms of warfare, information war (IW) will have offensive and defensive aspects. Militaries will both prosecute information war and defend against its use by the enemy.

2. Information war will be conducted at the strategic, operational, and tactical levels of war. Decision makers at each level will orchestrate information campaigns. They'll attack information infrastructures at the national, theater, and unit levels.

3. Information war will both support other military campaigns and operate independently. For example, just as naval air forces have both independent (e.g., antisubmarine warfare) and supporting (e.g., close air support) missions, information components will sometimes support other operations and sometimes require the support of other forces.

4. Information war will be an imperative for victory. Even as past victories were possible only through supremacy of the air, land, or sea, future victories will be doubtful without information supremacy.

5. Military forces must be capable of operations despite successful enemy IW. Because perfect defenses against IW are an unreasonable expectation, units must continue functioning despite corrupted information.

6. Information war will have distinct mission types. As with conventional military forces, no one type of IW will suffice to describe all its ramifications. For example, just as aerospace power has distinct mission types (e.g., airlift, interdiction, counterair), IW will have subsets. Table 2 illustrates this point in more detail.

Table 2

Comparisons Between Aerospace War and Information War

TYPICAL AEROSPACE MISSIONS	ANALOGS IN INFORMATION WAR
Counterair, Counter Space	Counter Information
Strategic Attack	Destroy/Distort National Information Network
Interdiction	Target C^4I Nodes
Close Air Support	Jam
Airlift	Transmit Information to Theater
Air Refueling	Update Databases in Flight
Electronic Combat	Insert Viruses, Corrupt Data
Surveillance & Reconnaissance	Understand Enemy's Information Architecture

Given the critical importance of information to future war, the theater commander should have a component commander

3

dedicated solely to winning the information campaign. Other components will have tactical information forces and interests to be sure, but to orchestrate information war at the strategic and operational levels, both offensively and defensively, the CINC should designate one commander and organization responsible for fighting and winning the information campaign.[5] This will be a critical campaign. Army Secretary Michael P. W. Stone reported after the 1991 Gulf War:

> The *first priority* of coalition forces during the offensive phase was the systematic destruction of Iraqi Command, Control, and Communications (C^3). The same offensive strategy is likely to be employed *against U.S. forces* by *any* future adversary. (Emphasis Added)[6]

The joint force information component commander (JFICC) should have five goals:

1. Collect information on enemy capabilities, deployments, and intentions.
2. Fuse data collected from all sources and distribute timely, filtered information to users.
3. Flow friendly information efficiently in the face of enemy attacks and competing friendly requirements.
4. Degrade enemy information networks.[7]
5. Defend friendly information networks against enemy intrusion.

To accomplish these goals, the JFICC should have operational control (OPCON) over certain forces on a routine basis, and should exercise temporary operational control over forces normally under the OPCON of other component commanders.

Forces normally under the JFICC's operational control should include *systems* (e.g., satellite communications [SATCOM], collection satellites, antisatellite [ASAT], specialized munitions) and *personnel* (e.g., satellite controllers, hackers, cryptographers, psychological operations [psyops] specialists, frequency managers). The JFICC should use assigned forces to conduct and support operations within the theater commander in chief's (CINC) overall guidance.

Caution: Whenever a new type of warfare emerges, there's a tendency to overstate its case. For example, during the 1920s and 1930s, airpower zealots overstated the capabilities of

4

airpower. Airpower visionaries, such as Douhet and Mitchell, made promises about the future effect of airpower that (to put it charitably) experience was slow to validate. The potential of strategic airpower was easier to foresee than to execute.

Information zealots will likely make similar mistakes. Because societies and militaries are increasingly dependent on information, the potential for information campaigns to fundamentally impact future war is obvious. However, its *practice* will take time to mature. This delay will be due to uneven progress in military reliance on information technologies, the ability to militarily affect information, and the military's acceptance of the attendant cultural changes.

For example, at the strategic level of war, information will have a decisive effect only if the target state is information-based. If it's a third wave state, its wealth will depend on information.[8] By targeting this information network, a military could impoverish its enemy and facilitate its defeat. However, if the enemy state's wealth is *not* based on information, a strategic information campaign will *not* have a decisive effect.

If the national leadership depends on an information network to retain power, this network should be a prime target. By severing the information network, the US could undermine the regime. But the network's value as a target set will decrease if the national leadership can maintain political power regardless of attacks on its information apparatus. Nevertheless, it will be important to attack the enemy's national information network due to its secondary effects on the nation's war-making (operational) capabilities even though these attacks won't be strategically decisive in their own right.

At the operational and tactical levels of war, however, planners should expect decisive effects from an information campaign. In 1994, the US Army Chief of Staff outlined this point.

> Information Age armies will develop a shared situational awareness based on common, up-to-date, near-complete friendly and enemy information distributed among all elements of a task force. First, operational and tactical commanders will know where their enemies are and are not. . . . Of course, this "knowledge" will never be absolute, and it is folly to assume it ever will become "perfect." It will be, however, of an order of magnitude better than that achieved even during the Gulf War. Second, information age armies will know where their own

5

forces are, much more accurately than before—and deny this critical information to the enemy. Last, this enemy and friendly information will be distributed among the forces of all dimensions—land, sea, air, and space—to create a common perception of the battlefield among the commanders and staffs of information age armies. [9]

If the US attains the Army Chief of Staff's vision—while degrading comparable enemy capabilities—it is difficult (though still possible) to envision our defeat. The speed, fidelity, and breadth of modern information systems offer orders of magnitude increases in military efficiency. This efficiency will only increase in the future. As a result, information efficiency will be a key factor in future war and will become an area of conflict. Commanders always seek to observe-orient-decide-act (the "OODA" loop)* faster than their opponents. Opposing fighter pilots, JFACCs, and national command authorities (NCA) always try to get inside their opponents' OODA loop. This was true in the past and will remain true in the future. The difference between the past and the future will be in terms of speed and breadth of decisions. As the technical ability to complete the OODA loop in near real time (NRT) emerges, commanders at all levels will move towards ever faster decisions. Whether it will be wise to do so is another question, but the ability to observe and command in NRT will exist—and whoever can get closest to it will gain an advantage.

This drive towards near-real-time C^2 will open interesting opportunities for operational art. Commanders will exploit their opponents' drives toward near-real-time decisions. Because near-real-time decisions will require heavy degrees of automation and decision protocols, commanders at all levels will strive to drive their opponent's snap decisions towards poor decisions, usually by presenting false indications of intent or reality.

For example, it is likely that US forces will be hypersensitive to any indications of missile TELs (mobile missile launchers) in the next war.** In fact, US C^2 may focus extensive attention on

---

*Term was coined by Col John Boyd.
**Besides TELs, other enemy systems which might attract considerable attention could include WMD and enemy leadership.

each and every indication of a possible TEL. If this is the case, and if the enemy identifies in advance which key indicators will immediately gain US attention, the enemy could quickly overload the US C^2 structure by providing "thousands" of single-factor indications of missile TELs. Single radio transmissions, heat bursts, shapes, or radar emissions might deceive US C^2 into rapidly refocusing its attention and surveillance assets on "thousands" of false targets. Even if each report demands only a few minutes of attention, this refocusing might affect all parts of US C^2 (not just the antimissile forces) as senior commanders get directly involved in sorting out the maze of false reports. To decrease the time required to deal with false reports, the US would probably either add more cross-checks into its TEL detection architecture (thus slowing the process) or restructure its decision algorithms regarding TELs (thus voiding some aspects of peacetime training). Either counterstep would drive major changes in US C^2 during the critical early stages of the war.

This contest over indications offers new possibilities in operational art. By understanding indication priorities and the tendency towards near-real-time ("snap") decisions, a commander could overwhelm the opponent's C^2 structure during the critical early phases of the fight.

The ultimate goal in this counter C^2 effort will be to compel the opposing force to either slow down its OODA loops or continue making bad decisions in near real time. Either option degrades information efficiency, thus gaining a decisive advantage in war.

Advances in hardware, software, and bandwidth—driven by the private sector—are certain. Their impact on future conflict will be profound. Simply stated, the ability to rapidly exploit observations of friendly and enemy positions and capabilities, at levels superior to that of the enemy, will be decisive at the operational and tactical levels of wars. For this reason, there will doubtless be a fight over information in any future war. Winning this information war—with integrated, redundant, secure, and exercised networks—will be imperative to victory.

Caution: The English language uses the word "know" to cover a broad range of sins. We can use "know" when referring to obvious facts (Do you *know* what time it is?). We can also use "know" when referring to in-depth assimilation of a subject (She really *knows* her business!). In English, the word "know" covers a broad range of cognition—from observation through understanding.

Unfortunately, this broad usage allows us to gloss over important distinctions. While it is certain that emerging information technologies will provide an incredible amount of detail about the enemy on day one of a war, it is less certain we will fully *understand* that enemy on day one. Satellites, UAVs, information "sweepers," manned aircraft, and civilian media will ensure we see the enemy with historic detail. However, simple observation of enemy deployments will not suffice; we'll also need an in-depth grasp of enemy forces at all levels of war. What is their logistics structure? Who are their key decision makers? How does their command and control function? What is their intelligence flow? Which units are reliable and highly trained/equipped? Which network nodes do they hold dear? What are their operational and strategic goals?

These questions, and thousands more like them, are critical to any attack plan. This plan demands extensive *under-standing* of the enemy, not just observation. When we blithely claim extensive knowledge about future enemies, we must admit this distinction. Which level of knowledge are we talking about? Without a concerted effort to study the enemy in advance of hostilities, our level of knowledge will probably bend more towards the observation side of the scale than the understanding side.

Parallel War

Although each section of this book treats mission areas separately, future aerospace operations against the enemy at all levels of war and across all target categories must occur almost simultaneously. Near-simultaneous attacks across the enemy target set will be the hallmark of future aerospace operations. Failure to conduct aggressive and overwhelming attacks across all facets of enemy power would waste a decisive capability.

The *theory* of near-simultaneous attack across multiple target sets is nothing new. Airmen have recognized it for decades. A large number of attacks in a day has far more effect than the same number of attacks spread over weeks or months. In his report to President Truman at the end of WWII, Gen Hap Arnold asserted that strategic air assault is wasted in sporadic attacks that allow the enemy to readjust or recuperate.

Historically, however, airmen lacked the military capabilities to implement near-simultaneous attack. During all of 1942–1943, for example, the Eighth Air Force attacked a total of only 124 distinct targets. [10] At this low attack rate (averaging six days between attacks), the Germans had ample time to repair and adapt between raids.

Contrast this WWII rate of attack with the 1991 Gulf War. In the first 24 hours of Operation Desert Storm, coalition air forces attacked 148 discrete targets. Fifty of these targets were attacked within the first 90 minutes. [11] Targets ranged from national command and control nodes (strategic) to key bridges (operational) to individual naval units (tactical). The goal was to cripple the entire system to the point it could no longer efficiently operate, and to do so at rates high enough that the Iraqis could not repair or adapt. Coalition forces, knowing an incredible amount about Iraq, efficiently orchestrated thousands of sorties, reached key vulnerabilities with high certainty, and, once in the target area, hit specific targets. The end result was near-simultaneous attack across hundreds of key Iraqi targets. Under this intense attack, Iraq was unable to either regain the initiative or orchestrate a cohesive defense.

Such targeting, conducted against the spectrum of targets in a compressed time period, is called *parallel war.** The goal of parallel war is to simultaneously attack enemy centers of gravity across all levels of war (strategic, operational, and tactical)—at rates faster than the enemy can repair and adapt. This is a new method of war. Previous generations of military strategists could not prosecute parallel war. They had only the sketchiest knowledge of the enemy's key strategic and operational targets. The enemy was opaque prior to contact.

---

*The term parallel war was coined by Col John A. Warden III, USAF. It came into general use after the Gulf War.

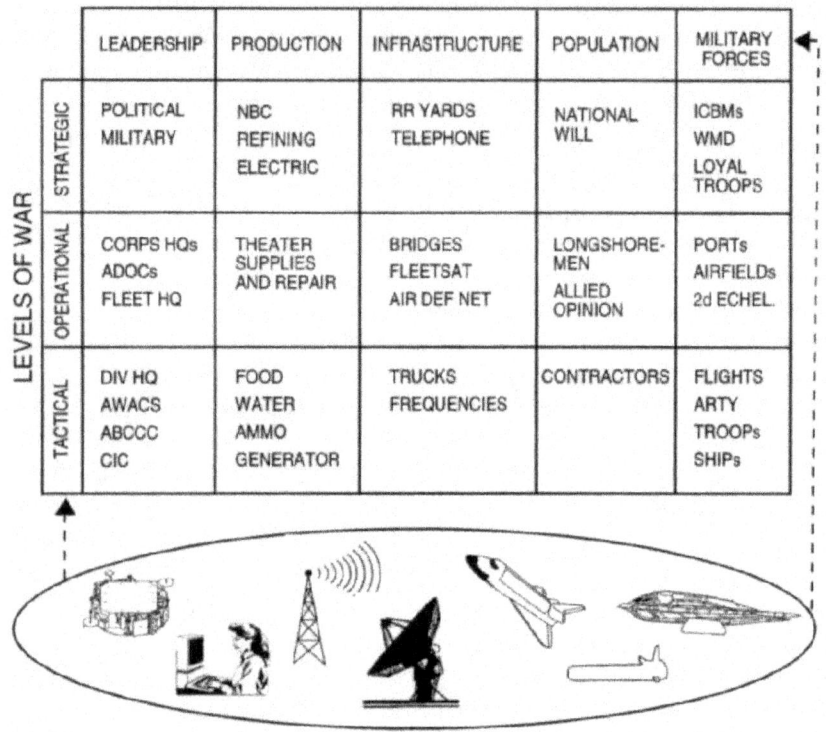

		LEADERSHIP	PRODUCTION	INFRASTRUCTURE	POPULATION	MILITARY FORCES
LEVELS OF WAR	STRATEGIC	POLITICAL MILITARY	NBC REFINING ELECTRIC	RR YARDS TELEPHONE	NATIONAL WILL	ICBMs WMD LOYAL TROOPS
	OPERATIONAL	CORPS HQs ADOCs FLEET HQ	THEATER SUPPLIES AND REPAIR	BRIDGES FLEETSAT AIR DEF NET	LONGSHORE-MEN ALLIED OPINION	PORTs AIRFIELDs 2d ECHEL.
	TACTICAL	DIV HQ AWACS ABCCC CIC	FOOD WATER AMMO GENERATOR	TRUCKS FREQUENCIES	CONTRACTORS	FLIGHTS ARTY TROOPs SHIPs

Figure 1. Target Categories

Even when military commanders knew what to target, they had to first "roll back" an enemy's defenses before attacking key centers of gravity. But modern technology is changing these long-held axioms of war. Extensive data on the enemy is now available on day one of any war. Although it will never be absolutely complete, the Information Age is providing ever increasing details on the strategic and operational centers of gravity of potential enemies. As demonstrated in the Gulf War, modern penetration and precision can place these centers of gravity under massive attack on day one of a war—and do so faster than an enemy can react. Most importantly, modern command and control systems can plan and direct this offensive in

CONFLICT	CEP	BOMBS REQUIRED*
WWII	3300 FEET	9070
KOREAN/VIETNAM	400 FEET	176
DESERT STORM	< 10 FEET	1
FUTURE	< 10 FEET (ALL WEATHER)	1

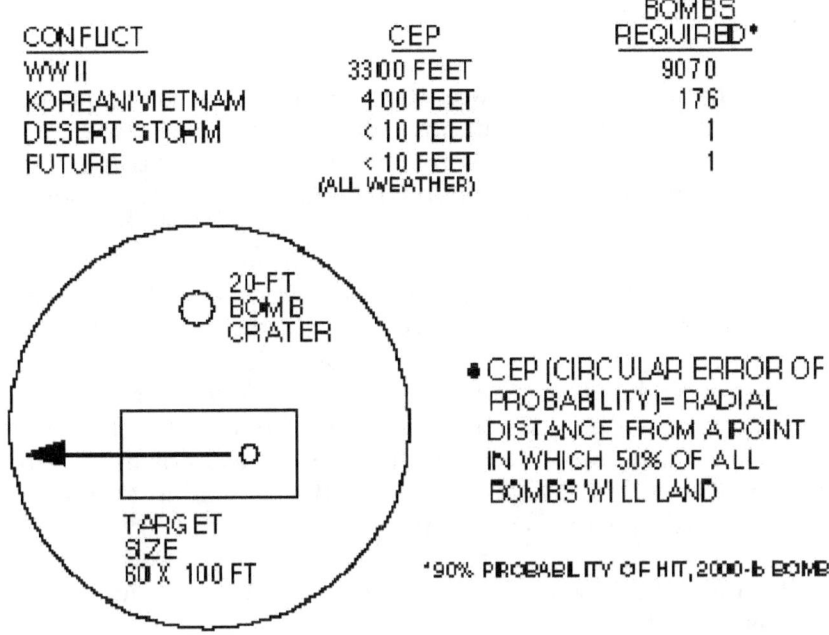

● CEP (CIRCULAR ERROR OF PROBABILITY)= RADIAL DISTANCE FROM A POINT IN WHICH 50% OF ALL BOMBS WILL LAND

*90% PROBABILITY OF HIT, 2000-b BOMB

Figure 2. Orders of Magnitude Improvement in Precision Munitions

near real time. These attributes of parallel war distinguish it from anything seen in military history. Ambassador Paul Nitze observed the effect of parallel war in the 1991 Gulf War:

> After the first few minutes, there was literally nothing Saddam could do to restore his military situation. His means of communicating with his subordinate commanders were being progressively reduced. The Iraqi commanders had difficulty carrying out orders for a coordinated movement of their forces. Their command posts, air shelters, and even tanks buried in the sand were vulnerable to elimination by precision penetrating bomb attack.*

*Paul Nitze, *Wall Street Journal* Editorial, 24 December 1991, 6.

Parallel war is enabled by emerging advances in *four key technologies:*

1. Information. By 2010, well into the Information Age, aerospace planners will detect an incredible amount of information about the target state. They will never know everything, but they will detect orders of magnitude more about the enemy than in past wars. At the strategic level of war, they should observe the connectivity among the national leadership, the architecture of the national communications grid, and the position of elite troops who are key to regime protection, among other things. At the operational level of war, they should see the location and connectivity of key corps and air defense headquarters, the naval order of battle, the location and LOCs of theater-level supplies, and the coordinates of critical nodes in airfields and ports. At the tactical level of war, they should know where most of the enemy's unit headquarters are, their communications centers and means, and the individual locations and readiness levels of squadrons, divisions, and ships.

2. Command & Control (C^2). Future commanders will use the Information Age's revolutionary advances in information transfer, storage, recognition, and filtering to orchestrate attacks and defenses. Theater-wide taskings will flow with unprecedented fidelity and speed. Commanders will convert "the understanding of the battlespace into missions and assignments designed to alter, control, and dominate that space."[12]

3. Penetration. Units will launch penetrating platforms against these targets. Enabled by stealth, hypersonic, and/or electronic warfare technologies, these platforms will penetrate in significant numbers. While defenses will certainly defeat some attackers, others will get through at rates higher than previously experienced.

4. Precision. Once over the target area, penetrating platforms will deliver brilliant munitions. Deliveries will be highly accurate. Target locations will be measured within feet. Circular error of probability (CEP) will be less than a meter. Brilliant sensors will have the ability to distinguish between tanks and trucks, between parked bombers and decoys. Because of this

precision, fixed and mobile targets will be struck by the thousands.

Attacks facilitated by advances in information, C^2, penetration, and precision will occur within the first 24 hours of conflict—and continually thereafter. This compressed, broad, and precise attack should leave the opponent paralyzed, unable to either coordinate an effective defense or mount an orchestrated offense. The potential for parallel war will only increase in the future. Information, C^2, penetration, and precision will allow targeting of each target type at the outset of hostilities. Advances in the underlying technologies will multiply the number of targets struck.

In 1995, the Air Force Chief of Staff described parallel war as a revolutionary development: "Not too far in the next century, we may be able to engage 1,500 targets within the first hour, if not the first minutes, of a conflict. . . . We will be able to envelop our adversary with the simultaneous applica- tion of air and space forces."[13] Unfortunately for the US, enemies will also have this capability. Employing—and defending against—parallel attack by aerospace forces will be a crucial aspect of future joint campaigns.

Revolution in Military Affairs

Modern warfare is in the midst of a revolution in military affairs (RMA). How US aerospace forces can thrive in this revolution is the guts of this book. RMAs are more than just changes in technology. Rather, RMAs occur only when militaries fundamentally change their concepts of operations (CONOPS) and their organizational structures to best employ radically new technologies. RMAs are *underwritten* by technology but *realized* through doctrinal change.[14] As the US Secretary of Defense noted in 1995:

> Historically, an RMA occurs when the incorporation of new technologies into military systems combines with innovative operational concepts and organizational adaptations to fundamentally alter the character and conduct of military operations.[15]

13

Throughout history, militaries have reacted differently to new technologies. Some opted to overlay new technologies on top of their current ways of doing business. They used new technologies to improve the efficiency of what they were already doing. Other militaries recognized the same new technologies as drivers of fundamental change. To realize the full benefit of the new technologies, they remade themselves; they remade their doctrine and their organization. In so doing, they gained substantial battlefield advantages over those who only overlaid new technologies on top of existing doctrine. A historical example illustrates this point.

During the interwar period, both Germany and France developed tanks. In terms of numbers and capabilities, their tank inventories were roughly the same. If anything, the French had a slight advantage in quantity and quality. In terms of doctrine and organization, however, the two armies were quite different. This difference proved decisive.

To realize the potential of the tank, the Germans remade their army. They recognized that mechanized armor rendered the old ways of war obsolete and, based on this recognition, the Germans did far more than add tanks to their inventory. Rather, they devised a new doctrine (*blitzkrieg*) and built new organizations to support that doctrine (*panzer* divisions). These were radical changes, and they fundamentally changed the way the German army made war. These changes affected promotions, strategy, maneuver, and training. The Germans also "bet their country" that it would work when they attacked France. It is interesting to note the Germans accomplished this revolution despite the severe restrictions imposed by the Versailles Treaty and the Great Depression.

The French army took a more conservative approach to tanks. They simply overlaid tank technology on top of existing CONOPS and organizations. They saw tanks only in terms of supporting infantry units. To an extent, they were correct—tank support made infantry units better. However, by using a radically new technology solely to improve and support existing units, the French missed the armored RMA of WWII. The result was Sedan 1940. The French had no idea of how to deal with the speed of German *panzer* thrusts, and they had no counter of their own. Although both armies had tanks (in fact, the French had more tanks than the Germans), their methods of employment differed greatly. *How* they employed this new technology was decisive. The fall of France provides a clear-cut lesson: CONOPS and supporting organizational structures are crucial when exploiting radically new technologies.

Today's aerospace planners face decisions of similar magnitude. Fundamentally new technologies are emerging. They will *underwrite* the next RMA. However, we won't realize the next RMA unless we devise new ways to employ the mix of emerging and present technologies, plus build organizational structures best suited to support this mix.

What are today's emerging technologies? There are four: information, C^2, penetration and precision. Future commanders will amass an incredible amount of data about the conflict arena, and they will have the technical means to cycle high-fidelity taskings in near real time. Weapons will reach targets throughout the depth and breadth of the theater and, after penetrating, these weapons will hit exactly where they're aimed. Previous generations of military leaders had bits and pieces of these capabilities, but they never had them all. The synergistic use of these technologies offers the potential for an RMA.

If history is any guide, aerospace forces must devise radically different CONOPS and supporting organizations to realize the full potential of the coming radically new technologies. It will be a singular stroke of luck if current US aerospace CONOPS and organizations bridge the gap between current and future technologies. Devising fundamentally new CONOPS and organizational structures will prove tremendously difficult, however. It will challenge career paths, hard-won modernization programs, professional military education, and a host of other facets crucial to success in war. Nevertheless, confronting this challenge is a prerequisite for realizing this revolution in military affairs.

Complicating our search is the fact that these technologies aren't secret. Both sides in a future war will have access to the same underwriting technologies. Both will have greatly improved information, C^2, penetration, and precision. Both may have innovative employment concepts and organizations. Therefore, planners must not only devise ways to use these new technologies; they must also make their operational concepts capable of succeeding while under attack from similar enemy capabilities. This is an immense challenge. It is the challenge this book is intended to evoke.

Caution: There is a natural tendency within today's military to focus on *defeating* these new technologies. We speak of information denial, viruses, antistealth radars, and spoofing technologies as having the potential to negate these emerging technologies. By orienting our defenses on these new types of threats, some argue, we can continue to rely on existing concepts of operations—concepts that have proven successful in past wars. Such thinking is a serious mistake.

When new military technologies arrive, we must learn to live with them. Hoping they'll go away is futile. Can you imagine army generals in the 1930s arguing that all they needed was an antitank weapon to make tanks obsolete? Or navy admirals arguing that all they needed was an antiaircraft gun to make airplanes obsolete? Unfortunately, historians tell us, some generals and admirals made exactly those arguments. They reacted to new—and unproved—technologies by focusing on defenses. To a degree they were right: Improved defenses made the tank and the aircraft vulnerable. Their error lay in equating vulnerability with obsolescence. The two aren't the same. The successful generals and admirals of WWII were those who exploited the capabilities and flaws in these new technologies while learning to live in the environment they created.

The successful generals and admirals of our next wars will be the ones who understand that advanced capabilities in information, C^2, penetration, and precision are here to stay. We can—and will—increase the vulnerabilities of these technologies, but we'll never make them obsolete. We must resist the temptation to believe that better defenses will allow us to return to the old and proven ways of doing business. Advanced information, C^2, penetration, and precision are integral to future war; we must adapt to thrive in their environment. This adaptation will realize the current revolution in military affairs.

Simulation

Today's state of the art is tomorrow's antique. Because of the rapid pace of information technologies, "cutting edge" today is "out-of-date" tomorrow. The *half-life* of today's hi-tech is measured in months, not years. If you doubt these axioms of modern life, ask anyone who has bought a personal comput er. Any computer bought today is invariably bettered in price, capacity, and speed within months. Whether we like it or not, rapid obsolescence in technology is the norm today. Staying on the leading edge of modern technology is a never-ending effort.

Nor is rapid obsolescence confined to the civil sector; military programs see the same effects. Computers in the first production J-STARS E-8C have 200 times the capability of those in the prototype E-8A deployed to the 1991 Gulf War . Subcomponents of the B-2 are already old technology. These examples are not criticisms of particular weapon systems; rather, they are examples of normal lag times between technology *introduction* and weapon system *production*.

It is important to stress, however, that procurement of a weapon system does not end this lag time. Putting an airplane on the ramp does *not* mean that it's ready for war. New weapons require trained personnel; operators and maintainers must learn how to use and fix their new equipment. New weapons also require integration into the rest of the force. Those charged with devising operational plans must thoroughly understand the capabilities and shortfalls of new systems. Supported and supporting organizations must also understand each new system's contributions. Realizing the full potential of a new technology requires retraining across the entire institution, and this takes time.

This time lag raises a critical issue for aerospace planners. Given a new technology, how much time can we spend in procuring the weapon and training the force—and still have a state-of-the-art weapon system? If procurement takes 10–15 years and training takes another 5–10 years, it is unlikely that state of the art will wait—unless no one else arms for war. That's a poor assumption for strategic planning!

In essence, today's rapid technological change means aerospace planners must compress *both* their procurement systems *and* their training systems. It does little good to compress one and not the other. It is important to field new technologies faster *and* to accelerate the training system. If operators must have a new system in their hands before they can put it in their brains, the cycle will always be long; it coul d induce a critical delay in the next war. The side that can best *identify* superior technologies and *train* its entire force to optimize them will have a decisive advantage. German Army doctrine emphasized this necessity over 60 years ago:

> New arms give ever new forms to combat. To foresee this technical evolution before it occurs, to judge well the influence of these new arms on battle, to employ them before others, is an essential condition for success.[16]

Achieving this advantage in an era of rapid change will require considerable innovation. One avenue having considerable promise for compressing the procurement training timeline is computer simulation. Computer simulation can contribute in five general areas:

1. Reduce the time, resources, and risks of the acquisition process. Simulation can also increase the quality of the system s being acquired. Virtual prototypes can support the many phases of the acquisition process and the testing of new systems.

2. Allow aerospace planners to develop doctrine and tactics while a weapon is still in development. Planners will be able to explore a new technology's incorporation into the complete spectrum of military operations prior to its fielding. Feedback can be available in near real time, with scenario recon-struction systems providing robust analysis capabilities. Information architectures, especially those dealing with automated C^2, can be tested across several mission areas.

3. Allow joint training with programmed weapons, active and reserve forces, multiple echelons, and large-scale forces. Computer-generated forces (friendly, neutral, and hostile) can replace human participants, allowing the representation of realistic large-scale forces in synthetic environments controlled by small numbers of human commanders. Synthetic environments can bridge large geographic regions worldwide and involve the entire joint force, from senior commanders down to individual airmen. Trainees can interact with synthe tic

environments through projected and alternative "go-to-war" command, control, communications, computers, and intelligence (C^{4}I) equipment and weapons.

4. Allow leaders to explore alternative plans, doctrines, and tactics. Computer simulation can support planners by providing insights into the effectiveness of theater-level campaign plans, operational-level battle plans, and tactical-level mission plans. Decision makers can simulate and evaluate consequences of alternative courses of action. Automated scenario generation and database construction tools, along with easily accessible database repositories, can enable short-notice models and simulations, allowing computer simulation to support crisis action planning.

5. Allow war fighters and military planners to rehearse missions. Simulators can immerse war fighters in a synthetic environment that accurately models the anticipated terrain, environmental conditions, and threat. This capability can increase the probability of mission success by fostering familiarity and proficiency with the mission plan while providing feedback to improve the plan.

Simulation is only one part of the overall acquisition and training process. It has definite limitations, not the least of which is its poor capabilities at the strategic level of war. However, computer simulation has the potential to aid both analysis (making better decisions) and training (inducing better behavior). It can be conducted with varying levels of human interaction. Its greatest attribute may reside in identifying interfaces and analyzing interoperability. Varying aspects of warfare (joint, ground, air, naval, NBC, R&D, training, C^2, communications, logistics, intelligence, space, special operations, etc.) are all amenable to some measure of computer simulation.

Staying on the cutting edge of rapidly advancing technologies is tough to do. It requires continual learning and reequip- ping. To compete in this environment demands an ability to quickly procure and incorporate new hardware and software. Failure to do so will result in CONOPS and equipment that fall short of state of the art. Computer simulation has considerable potential for helping us to thrive in this era of rapid change.

19

Notes

1. Adm William A. Owens, "Vision Force 2005: The Pending Revolution," briefing, vice chairman joint chiefs of staff, March 1995.

2. Inserting false data into an enemy's information system is probably the most effective tactic in information war. It encourages the enemy to: (1) mistrust all its data; and (2) construct internal barriers (or gateways) for data entry.

3. Foreign Broadcast Information Service (FBIS)-CHI-93-126, 2 July 1993, 22.

4. Vladimir I. Slipchenko, "A Russian Analysis of Warfare Leading to the Sixth Generation," *Field Artillery*, October 1993, 38.

5. In a similar vein, all components employ aircraft, but only one (the Air Force) has the primary missions of strategic attack and defeat of the enemy's air force.

6. Secretary of the Army Michael P. W. Stone, *Preliminary Report. Lessons from Operations Desert Shield and Storm* , 22 April 1991, 7. Entire report is classified (Secret). Information extracted is unclassified.

7. Richard O. Hundley and Eugene C. Gritton, *Future Technology-Driven Revolutions in Military Affairs* (Santa Monica, Calif.: RAND, 1994), 68.

8. Alvin and Heidi Toffler, *War and Anti-War: Survival at the Dawn of the 21st Century* (Boston: Little Brown, and Co., 1993). *First Wave* states are agrarian-based. *Second Wave* states are industrial-based. *Third Wave* states are information-based.

9. Gordon R. Sullivan and James M. Dubik, "War in the Information Age," *Military Review*, April 1994, 56.

10. Roger A. Freeman, *Mighty Eighth War Diary* (London: Jane's Publishing Co., 1981), 9–161.

11. CENTAF Master Attack Plan, 16 January 1991. Although the plan is classified, these figures are unclassified.

12. Adm William A. Owens, "The Emerging System of Systems," *Proceedings*, May 1995, 38.

13. Gen Ronald R. Fogleman, "Getting the Air Force into the 21st Century," speech to the Air Force Association's Air Warfare Symposium, Ireland, Fla., 24 February 1995. Italics and underlining in the original.

14. For a more complete overview see James R. FitzSimonds and Jan M. van Tol, "Revolutions in Military Affairs," *Joint Force Quarterly*, Spring 1994, 24–31.

15. William J. Perry, Annual Report to the President and the Congress (Washington, D.C.: Department of Defense, 1995), 107.

16. Excerpt from *Die Truppenfuhrung*, 1933, quoted by Matthew Cooper, *The German Army 1933–1945: Its Political and Military Failure* (New York: Stein and Day, 1978), 137.

Chapter 2

Peer Competitor

Along with today's military focus on major regional contingencies (MRC),* moderate-sized conflicts, airmen must also prepare for war with a future peer. While major regional contingencies are more likely and deserve immediate attention, planning for them won't suffice for war with a peer. War with a peer would involve higher degrees of scope, casualty tolerance, national mobilization, and—most importantly—enemy military capabilities. In a war with twenty-first century equivalents of Nazi Germany or the Soviet Union, the risks and downsides of a US defeat would be great—far greater than in an MRC-level war. To twist a cold war phrase, war with a peer would not be a "greater included case" of an MRC.

A peer competitor is capable of fielding multiple types and robust numbers of both emerging and present weapons, then developing a new concept of operations (CONOPS) to realize the full potential of this mix. Its goal is to capture a vital interest of the United States, then defeat the US military response.

Fortunately, the chance of war with a peer is remote. The US has unquestioned military superiority over all possible adversaries. No potential peer nation is arming for war with the United States. The US currently exceeds every defense budget in the world by at least a factor of four, spending as much on defense as the next eight largest defense budgets in the world combined.[1] The US is at peace with the few nations capable of reaching peer status; war with a peer is not on the horizon.

Unfortunately, this favorable environment won't last. If history teaches us anything, it teaches us times change. Despite current optimism, humankind has *not* seen the end of major war. Major war may happen in 10 years (unlikely),

*The term Major Regional Contingency (MRC) was defined by US Secretary of Defense Les Aspin in *Report of the Bottom-Up Review*, October 1993.

21

or 15 years (possible), or sometime after that (virtually certain). Defense planners should regard conflict with a peer as inevitable; only the timing is unknown. For discussion purposes, this book assumes the early edge of the window—it discusses war with a peer beginning in 2010.

This 2010 time frame is arguable. Some may view 2010 as too far away; they believe it a waste of time to fine-tune future decisions because we'll revisit today's decisions many times between now and 2010. On the other hand, others consider 2010 too close; because it's only 15 years away, they say, most of our present inventory will still exist. In fact, they point out, some systems in the current program objectives memorandum (POM) will still be in production in 2010. They recommend that we really stretch our thinking and consider a war in 2020 or 2030. To clarify the reason for a 2010 focus, it might be helpful to draw an analogy with the interwar period.

In the mid-1920s, the "war to end all wars" had just ended. Democracies were triumphant. International tensions were few. Military spending was tight. Few people thought another war between the great powers was probable. In fact, most people considered such a war inconceivable. Unfortunately, however, World War II was only 15 years away. Great power conflicts sparked by unforeseeable events initiated rearmament in the mid-1930s—and global war five years later. In this short period of 15 years, international politics radically changed. Linear projections of the future became worthless. Conflict with a peer went from being inconceivable in 1925 to the national purpose 15 years later.

It is important to understand that the rapid change of the interwar period was not unique. Swift, discontinuous change is the norm throughout modern history. The normal course of events is to experience fundamental change in short periods of time. However uncomfortable this history is for planners, we must recognize rapid change in international politics and technology is a given. The change witnessed between the Treaty of Locarno (1925) and the Battle of Sedan (1940) reflected many other 15-year periods in history. For example, consider the 15 years starting in 1945. Imagine giving a speech in 1945 in which you predicted that, by 1960, the US would

- begin 40+ years of "cold war" with the USSR,
- see the Soviets test an A-bomb before the end of the decade,
- see the USSR launch the first space satellite,
- see China "fall,"
- fight a communist China in Korea,
- strongly back the new Jewish state in Palestine,
- build atomic-powered ships, and
- face a communist Cuba.

You would have been laughed off the stage. Your audience would have rejected the possibility of these changes. Nonetheless, each event occurred, with profound impacts on the US military. Policymakers had to face a radically different strategic situation. Another example would be the next 15-year period (1960–1975). The US lost a war, witnessed a president assassinated, saw a president and a vice president resign in disgrace, watched as France withdrew from NATO's military structure, began the "Great Society" program, went off the gold standard, and saw one-dollar-a-gallon gasoline. If you're still not convinced, take the next 15-year period, 1975–1990. Who in 1975 predicted seeing, within 15 years:

- Computers on every desk
- The US as the world's number one debtor
- Quarter-trillion-dollar federal budget deficits
- Honda as the number one car in America
- Capitalism in China
- Disintegration of the USSR
- A US-Iraq War (with Russian approval and Syria as an ally)
- Peace between Egypt and Israel
- The Iranian Revolution
- A Polish Pope

As we look towards the future, we should keep these lessons of modern history in mind. The world can change radically in as little as 15 years—and usually does. If we are lucky, change over the next 15 years will be positive and peaceful. We may experience an extended period of "deep peace." Statesmen may successfully avoid war. Nations may concentrate on internal and economic matters. Whatever change we experience may be peaceful and stable. In fact, we can make a good case "deep peace" is the *most likely* future scenario. It's tough to sculpt a

23

credible scenario where great powers have no better option than war. None of the great powers should fight over farmland (food is plentiful and cheap). Nor should they fight over resources. Oil, the only obvious resource that might force a war, should remain abundant for the next several decades.* When building war games, our planners have great difficulty devising a credible cause for major war. Major war is difficult to posit as a likely scenario.

Nevertheless, prudent planners cannot dismiss war's possibility. The summer of 1914, which saw an assassination turn into a World War, provides sufficient precedent for political stupidity. Whether we like it or not, great powers will eventually "stupid" themselves into war. This war could start after some sort of crisis. Its nature could be environmental, viral, economic, or ideological. Regardless of cause, we must assume a major war will someday happen. Miscalculations by one or more of the parties will eventually escalate a crisis into a major war. [2]

With this possibility in mind we must add another variable. Just as the international environment can change quickly, so can military operations. Again, the interwar period is instructive. In the short period between 1925 and 1940, the science of war changed radically. Armies, navies, amphibious forces, and air forces underwent revolutions. The German army developed armored warfare. The US and Japanese navies developed carrier warfare. The US Marines developed amphibious warfare. The Royal Air Force and the US Army developed aerial warfare. Advances in technology and doctrine revolutionized warfare. As a result, the character of war fought in 1940 was entirely different from that possible in 1925.

Changes in air force inventories illustrate this point. In 1925, biplane bombers such as the Curtiss B-2 were state of the art—and extremely expensive. Within 15 years they were obsolete. Aviation technology sped past the biplane. The heavy bombers and monoplane fighters of World War II, with their training, logistics, and basing infrastructures, bore little resemblance to the inventory of the Army Air Service in 1925.

---

*Some people include "fresh water" as a future scarce resource. However, given enough oil, a state can convert sea water to fresh water. Other critical resources have either substitutes or multiple sources.

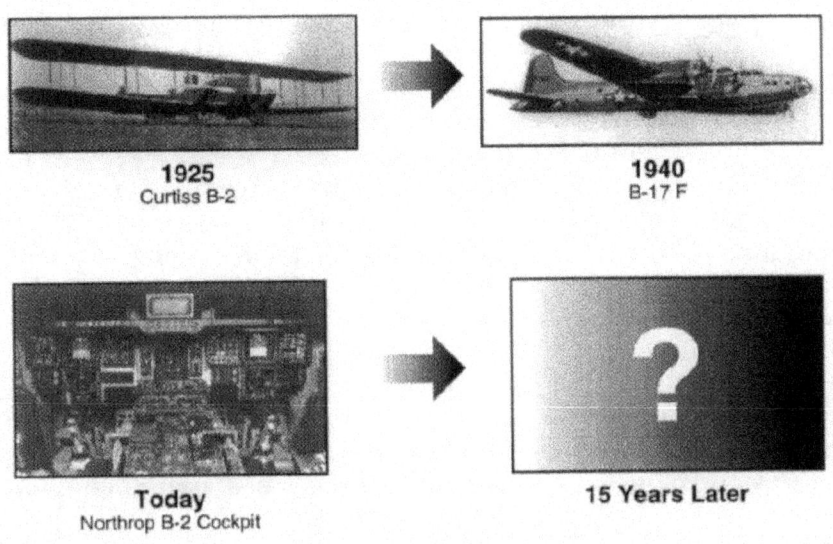

Figure 3. Radical Change in 15 Years

Although the equipment of 1925 quickly became obsolete, the thinking of this era directly affected initial WWII operations. During the interwar years, prophets like William ("Billy") Mitchell and groups like the Air Corps Tactical School built the doctrine of strategic bombing. Their thinking, combined with the civilian aircraft industry's emphasis on large aircraft, drove the Air Corps focus on heavy bombers. As a direct consequence of this focus, most of the funding during the tight budget years of the Great Depression went to heavy bombers. This procurement made B-17s available at WWII's outset; B-17s embodied the Air Corps doctrine of strategic bombardment.

At the same time, this doctrine dampened development of long-range fighters. For this reason, large numbers of P-51s and P-38s did not arrive in Europe until two years after the war began. Even when these long-range escort fighters did arrive, it took several months of trial and error to devise their optimum CONOPS.[3] This shortfall in long-range fighter inventories and CONOPS was a direct result of interwar thinking.

25

During this short period, both the inventories and the doctrine of airpower underwent fundamental change. Of these two changes, developments in doctrine were far more enduring. While the inventories of 1920–1935 were irrelevant to WWII, the doctrines developed during that era dictated the initial employment of airpower in WWII.

Although major conflict is currently improbable, given current international conditions, history tells us that those conditions will quickly—and radically—change. For reasons unknowable today, major conflict will erupt at some time in the future. For planning purposes, we must assume (1) some sort of major crisis will eventually heighten international tensions; (2) these tensions will spark a military buildup, and (3) war will follow.

Could all this happen within the next decade? Very unlikely. The stream of events required to produce a major conflict between great powers will take time. Only when we project beyond 15 years do we enter the realm of the possible. Unlikely to be sure, but still possible at the early edge of this window. Given the downsides inherent in being unprepared, we believe it necessary to explore this possibility.

To envision this future war, planners should start with possible future weapon systems as their baseline— not what is currently on the ramp and in procurement. As the WWII experience shows, most of today's weapons will be obsolete for a 2010 war. For example, it is very unlikely that today's models of cruise missiles and satellites will reflect the state of the art in 2010. Nor will bombers. Just as advances in engine technologies made the 1925 Curtiss B-2 bomber obsolete in WWII, advances in information technologies will bypass the avionics/computers/munitions in today's Northrop B-2 bomber.

Although today's weapons will become obsolete, today's thinking will not. The doctrines developed today will be critical. If the World War II analogy holds, doctrines developed today will guide rearmament and initial operations in the next war. Today's planners will develop the operational concepts for a 2010 war; how US aerospace forces *fight tomorrow* will be guided by how US aerospace planners *think today*. For this reason, we need to explore the concept of war with a peer competitor in the 2010 time frame.

Environment

When projecting a major conflict with a peer, planners must expect both sides to employ significant numbers of advanced-technology aerospace systems. These systems will include

1. Atmospheric and space-based reconnaissance and communications systems. These systems will vary in quality and quantity between opponents. They will, at a minimum, be able to detect massive force movements and relay this information in near real time despite significant enemy countermeasures.

2. Information Age command and control systems. Future C^2 will devise and direct integrated taskings with high fidelity in near real time. They'll be heavily automated and dispersed. Attacks on any single node of this structure will not have catastrophic effects.

3. Stealth aircraft and stealth cruise missiles. Both sides will deploy tens of thousands of aerospace weapons with low signatures. These very low-observable weapons will use state-of-the-art electronic warfare systems to further increase their chances of penetration. Stealthy cruise missiles will be inexpensive, allowing their employment in massive numbers.

4. Precision weapons. Reflecting current trends in sensor technologies, precision weapons will have less than one meter accuracy with brilliant munitions.[4] They will guide independently of external positioning systems (e.g., global positioning system [GPS]), and they will have automatic target recognition capabilities.[5] Some of these weapons will retain their accuracy regardless of weather or darkness.

In addition to these emerging technologies, both sides will possess large numbers of nuclear weapons plus delivery systems capable of worldwide reach.* This strategic nuclear threat will significantly constrain military operations. Due to the possibility of nuclear retaliation, each side may place restrictions on attacks against the other's homeland. Political leaderships may prohibit attacks on certain strategic targets (e.g., leadership, satellite ground stations, enemy stealth

---

*This book does *not* assume international nuclear disarmament.

facilities) within the enemy's borders, regardless of means. Both sides will have substantial resources and strategic depth. Neither will be overwhelmed by sheer numbers. Both will have an economy capable of producing large numbers of state-of-the-art weapons. Both will have enough territory to permit maneuver.*

War with a future peer will present challenges of a different nature from those posed by an MRC scenario today. Both sides will use multiple sensors to detect large force movements and relay this information in near real time to stealthy aerospace weapon systems. Possibly operating from a sanctuary, these stealthy aerospace weapons will likely penetrate aerospace defenses in significant numbers. Once in the target area, they will strike with great accuracy. Most importantly, these weapons will be employed and controlled in an innovative fashion. Both sides will employ emerging technologies in ways that maximize their unique capabilities. Defense forces will face a combination of advanced surveillance and communications, innovative command and control, stealthy attack systems, precision munitions, nuclear weapons, and robust resources in the hands of an innovative attacker.

The fact that this war-fighting environment will be challenging and destructive does *not* mean US aerospace forces can't surmount it. Quite the contrary. If the US plays its cards right, it could thrive in this environment. The US already possesses early generations of the key emerging technologies. For example, the US is experimenting with fourth generation stealth aircraft while other nations are still trying to understand stealth's basic physics. Stealth ownership allows the US to devise counters and improvements to stealth in practice while others must rely on theory. In addition to stealth, the US leads most potential enemies in precision weapons, space platforms, all-weather enabling technologies, information war, and simulation. As a result of this head start,

---

*A precise definition of a peer in terms of size and depth isn't possible. If size is a valid criterion, it seems safe to say that small countries, along the lines of Singapore and Israel, could never attain peer status. Countries with the size and wealth of Korea , Brazil, and South Africa could.

the US can refine and integrate a series of key emerging technologies while other militaries are still trying to build them.

In any war with a peer, US force structure would be far greater than currently programmed. We can anticipate substantial US rearmament as a peer competitor arms itself and relations sour. As that happens, the US will draw upon its resources to rapidly field large numbers of the latest generation of weapons. The US will not "sit pat" while a potential peer enemy arms for war.

But technology alone doesn't win wars; *operational art* is decisive. Given rough parity in weaponry, whoever best employs its weapons wins the battle. History is replete with examples where both sides employed roughly equal forces but with quite different employment schemes—and completely different results. Sedan (1940), Midway (1942), and the Bekaa Valley (1982) are but three examples of campaigns in which the victor used a superior concept of operations to over-whelming advantage. As in past wars, future battles will be won by the side that has the best concept of operations.

Today's aerospace planners must devise superior employ-ment concepts for future weapons. Given the US lead in technology and resources, the US should have superior weaponry in a war with a peer. Whether the US will have a superior CONOPS is less certain. In building a future CONOPS, planners should start by forecasting future weapons capabilities for the US and its peers. They should then ask themselves whether current US offensive and defensive CONOPS will thrive in that environment. If the answer is yes, our problem is greatly simplified; planners need only incorporate these new weapons into current plans. If the answer is no, however, planners must build new US offensive and defensive aerospace CONOPS.

As a first step, we should ask ourselves: Will the current US air *defensive* CONOPS suffice against a peer in 2010? Unfortunately, the answer is probably "no."

The current *air defense CONOPS* for all American military forces assumes beyond visual range (BVR) detection of enemy aircraft and missiles. We assume that long-range sensors, primarily radar and infrared, will detect and track enemy aircraft and missiles far from the target area. Given this long warning time, air defense C^2 will have time to sculpt a response. We further assume that commanders will have sufficient time to pick the most efficient weapon, task that weapon in a positive manner, and perform cross-checks to decrease the chance of fratricide. For example, current US weapon systems are built on the assumption that long-range sensors will acquire the target. Patriot, AIM-7, AIM-120, and AWACS assume that the target has a high radar signature; DSP and AIM-9 assume that the target has a high infrared signature. Thus, a key assumption throughout the current

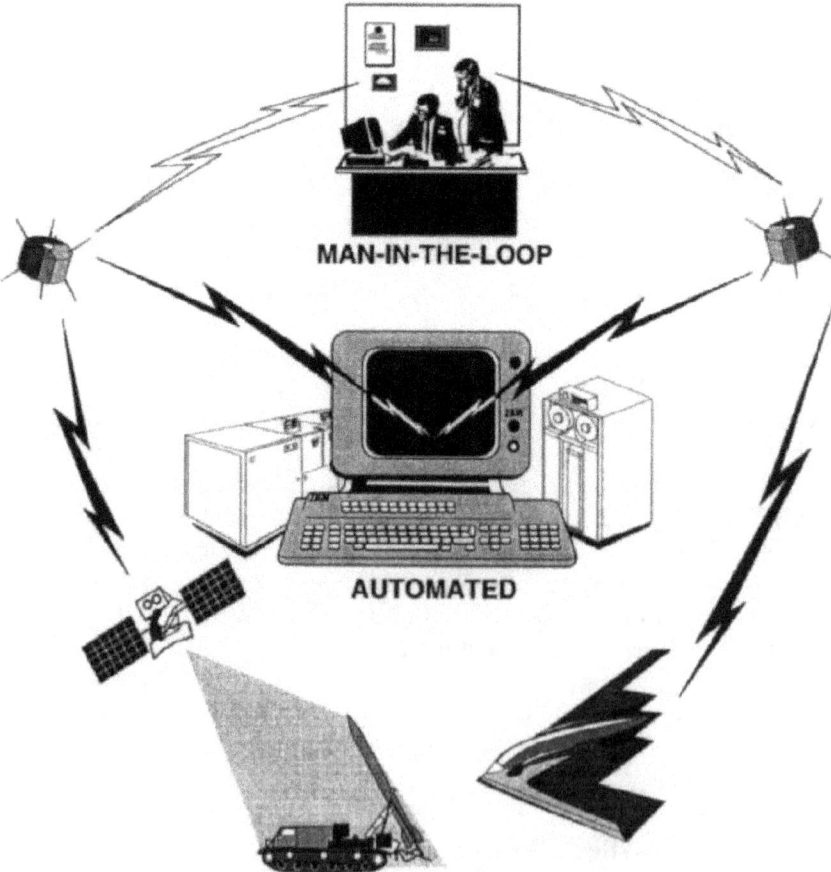

MAN-IN-THE-LOOP

AUTOMATED

Figure 4. Automated Versus Man-in-the Loop Decision Making

30

aerospace control CONOPS is that enemy aerospace platforms will reflect or emit high signatures.

Unfortunately, that assumption will not prevail; future warfare will involve thousands of stealthy cruise missiles and aircraft with *low* signatures. The heat signatures of aircraft and cruise missiles will be below the tolerances of spaced-based infrared surveillance systems, making them difficult to detect upon launch. Stealth technologies will decrease their chance of detection by radar. In addition, aircraft and cruise missiles will avoid intense defenses by varying their routes. Even if detected in flight, their target will be uncertain. For all of these reasons, alerting specific terminal defenders will be difficult.

Our present CONOPS also assumes limited numbers of attacking missiles and aircraft. Due to the multimillion dollar unit costs of aircraft and accurate ballistic missiles, we can assume that any attack by these systems will be limited. For example, the entire US Air Force (active duty, guard, and reserve) inventory totals only 6,814 aircraft.[6] While large in a relative sense, this number is small in an absolute sense. A limited inventory means limited attacks. Attacks can involve only a few hundred at a time; at most a thousand. Reflecting this limitation, coalition air forces launched only 931 attack sorties during the first 24 hours of Operation Desert Storm.[7] Given these relatively limited numbers, the current aerospace defense CONOPS is appropriate. A few hundred costly attackers justifies multiple defensive shots by less expensive (but still costly) SAMs and AIMs. Stealthy cruise missiles, however, change this exchange ratio.

Stealthy cruise missile s are cheap. One US defense contractor reported his company could build a –30db (front and rear aspect) cruise missile with 300 NM range for $100,000. He then added that one should *not* buy this missile from his company; a company with less overhead could build the same missile much cheaper.[8] Expected advances in production technologies combined with economies of scale (driven by large procurement runs) should cut the costs of very low-observable cruise missiles even further.

Such low unit costs will allow a peer attacker to employ stealthy cruise missiles in waves. At $100,000 per copy, a fleet

Figure 5. Massive Stealthy Cruise Missile Attack

of 100,000 stealthy cruise missiles would cost only $10 billion. Such a sum is well within the range of any peer anticipating war with the US. A fleet this size could launch waves of attackers. Each cruise missile would be cheaper than the US defensive weapon sent against it (SAMs, AIMs). The current US aerospace defense CONOPS, which shoots expensive missiles at even more expensive aircraft and ballistic missiles, is ill-suited to a massive, continuous attack by cheap cruise missiles .

Another factor that must be considered is *command and control*. Current C^2 concepts for US aerospace defense are ill-suited to the emerging environment. With few exceptions (e.g., Patriot batteries in automatic mode against incoming missiles), lethal attacks on aerospace targets require human decisions. Human fingers control every trigger. Usually, voice commands are required prior to missile launch. In an era of multiple penetrating targets, each with low signatures, such positive control may prove insufficiently responsive. Only an automated C^2 structure will have the speed to react in sufficient time to defeat a mass attack by low-signature missiles. Unfortunately, the culture of current aerospace organizations will slow the understanding of this shortfall.

Another C^2 shortfall is in the area of *doctrine*. US adherence to the doctrine of decentralized execution[9] will degrade defensive operations. Because of the increasing range of defense weapons, multiple defenders may fire on the same target at the same time. They may all have the motivation and opportunity to engage the same target simultaneously. Different batteries of SAMs and flights of interceptors may also overlap coverage of specific targets. We need to deconflict firing decisions across our broad array of defensive weapons in this environment.

Given a fast, lethal, and low-signature target, several defenders may feel the need to quickly take any shot that presents itself. Given decentralized C^2, several aircraft/batteries might fire on the same target simultaneously. Or one platform might shoot while another makes a counter-productive maneuver. Or *no one* might shoot, each thinking that another defender has the lead. The most appropriate defender may even withhold its fire due to fears of threats yet to appear. Low-signature targets pose a considerable problem for future defenders.

Stringent rules of engagement (ROE) may solve the deconfliction problem *if* all possible circumstances are worked out in advance of the war. But such prescience is unrealistic. It would require an accurate projection of enemy capabilities and friendly vulnerabilities in advance of the war. There's no historical precedent for such an accurate projection. Therefore, an alternative is necessary. Only a centralized C^2 system has the potential to deconflict these factors in the chaos of war. Directing long-range defensive missiles against short-range targets presents an immense C^2 challenge. Decentralized execution, effective in past wars, won't answer this challenge.

For all of these reasons, sophisticated stealth in the hands of a peer enemy would render our current aerospace defense CONOPS *obsolete*. If the US attempts to use its current air defense CONOPS against a future peer aerospace threat, it would *not* be able to enforce air supremacy.[10] Stealthy attackers would likely penetrate in high numbers. Taking advantage of modern surveillance and precision, they would hit crucial targets with substantial effect. In essence, high leverage enemy air attacks against US deployments would be probable. Therefore, the US needs a new aerospace defense CONOPS to survive in this future environment.

In this same context, we must also review future *offensive* operations. Will the current US offensive CONOPS suffice in the future? Unfortunately, the answer again seems to be "no."

The current US aerospace CONOPS anticipates extensive use of in-theater systems. The overwhelming majority of these in-theater systems (e.g., AWACS, KC-10, F/A-18E, F-15E, Army Tactical Missile System [ATACMS]) emit or reflect high signatures. If employed against a future peer, they would be highly vulnerable to detection by multiple layers of enemy sensors. With this information, the peer enemy will inflict substantial attrition. Stealthy interceptors (whether manned or unmanned) will attack the airborne platforms. Stealthy cruise missiles and bombers will attack their bases. Short-legged US stealthy systems, such as TLAM, F-22, and F-117, would also be vulnerable. While survivable in flight, they depend on high-signature support systems (e.g., surface ships, AWACS, air refuelers, fixed air bases). By attacking these high-signature support systems, a peer enemy could significantly degrade short-legged US stealth. These vulnerabilities point to a recurrent theme in future warfare theory: high-signature systems won't survive. This theme applies to aerospace forces as well as their ground and naval cousins.

The stealthy cruise missile symbolizes this threat. Future stealthy cruise missiles will (1) fly against critical targets, (2) penetrate into target areas in large numbers, and (3) hit within feet of their targets. Stealthy cruise missiles, properly supported by information and precision technologies, will make high-signature, immobile forces extraordinarily vulnerable. Air supremacy, which is required to protect ammunition ships needing days to unload, airlift aircraft needing hours to off-load and refuel, large air bases with "tent cities," and air refueling aircraft parked nose-to-tail in the open, may not be possible. In this emerging environment, the United States may not be able to protect high-signature, theater-based aerospace forces. Absent a reasonable certainty of protection, any CONOPS dependent on their survival is suspect.

Simply stated, we'in which the current US aerospace CONOPS will prove inadequate for dealing with an enemy employing advanced information, C^2, penetration, and precision in a sophisticated manner.

In addition to its impact on war fighting, this vulnerability also raises a stability issue. Lacking an ability to absorb an enemy attack in this new environment of advanced information, C^2, penetration, and precision, we will be tempted by an overwhelming incentive to preemptively attack. This defines a dangerous situation: Absent adequate defenses, whoever strikes first wins. When either side in a crisis perceives an overwhelming advantage by striking first, that crisis will be inherently unstable.

Nuclear deterrence doctrine during the cold war addressed crisis stability in great depth. To induce crisis stability, both sides built large inventories, redundant systems (e.g., the TRIAD), extensive surveillance, hardening, and innovative CONOPS (e.g., airborne alert). The intent of these measures was to heighten crisis stability. The nuclear deterrence theorists did more than envision how to fight a nuclear war; they also described how to avoid "use or lose" situations. Today's aerospace planners must sculpt similarly effective crisis stability regimes for the emerging stealth environment.

In summary, today's aerospace planners must devise a future aerospace CONOPS with three projections in mind. First, aerospace defenses must anticipate a massive, low-signature target set. CONOPS that assume long-range detection of limited attackers will not thrive. Second, offensive aerospace forces must de-emphasize high-signature, theater-based forces. Their attrition in the emerging environment will be sufficiently high to preclude high-tempo operations. Third, planners must take steps to induce greater crisis stability into the US force structure and CONOPS. Absent greater redundancy and more effective defenses, the US could find itself in a "use or lose" predicament during a crisis.

With these three themes in mind, the following 11 operational concepts will be critical to aerospace operations in a future war with a peer.

- Conduct a defensive counterstealth campaign.
- Degrade enemy cruise missile guidance.

- Establish ballistic missile defense.
- Control and exploit space.
- Integrate ISR (intelligence, surveillance, reconnaissance) systems.
  - Support the Information Campaign.
  - Conduct offensive strikes within enemy homeland.
  - Attack enemy invasion/occupation forces.
  - Avoid deployment of critical targets within range of enemy stealth.
  - Position JFACC in CONUS.
  - Airlift critical supplies and spare parts into the combat area.

Conduct a Defensive Counterstealth Campaign

"Stealth" is synonymous with low observability—not invisibility. Stealth systems will reflect or emit signals intermittently during flight. Their "stealthiness" will vary depending on aspect. While their frontal aspect may present a very low-observable signature (–25 to –30 db), their side, rear, or overhead aspects may reflect to a much higher degree. Different radar bands will also offer different levels of detection. By thoroughly fusing different types of sensors throughout the battlespace, defenses might increase their detection of stealth systems—thus enhancing the cues available for friendly fighters and air defense batteries. Once cued, those defense systems could focus their sensors on a specific area to track and target. The mix of defense sensors should include these six characteristics:

1. Long-wavelength radars
   a. Land-based (due to power requirements)
   b. Airborne (to expand line-of-sight range)
2. High-altitude, possibly space-based, radars (to give a vertical aspect)
3. UAVs with radar, infrared, and imaging sensors
4. SIGINT (to cue airfield attack, detect sortie generation)
5. High-power, short-range radars/lidars* arranged along likely attack corridors (mobile to degrade preemption efforts)
6. High-fidelity and near-real-time kill assessment

---

*Lidar = light detection and ranging (a laser radar).

Sensor deployment must keep three principles in mind. First, all sensors must feed an integrated database. Stealth systems will not allow many "hits." The few detections received might give a targeting solution if thoroughly fused. A fused system of sensors might also decrease the chance of enemy spoofing. Second, these sensors should be arranged in a circular, vice linear, fashion. Stealth platforms have varied signatures based upon aspect. It will be far easier to detect a stealth missile/aircraft from the side or rear than from head-on. In addition, because stealth platforms attempt to reflect radar energy away from the transmitter, it would be a great advantage if radar reflections from pulses emitted from one location could be received in a second location. Third, as the enemy tries to spoof sensors or as expected signatures fail to mirror reality, field units must have the capacity to rapidly adjust sensor algorithms.[11] It will do little good to identify every inbound stealthy target if multiple false targets are concurrently displayed. The sensors must be designed to allow rapid adjustment by trained operators.

Once wide-area defense sensors detect stealth missiles/ aircraft, they will cue air defense interceptors and missiles. The interceptors will need the following nine characteristics:*

1. Data link with the sensor fusion center. Onboard sensors will not suffice to acquire and target enemy stealth. Interceptors will need real-time updates from offboard sensors.

2. Long range. Interceptor bases should be beyond the range of enemy cruise missiles (as should aerial refueling bases). Interceptors must sortie from rear bases and loiter in the expected engagement area.

3. Air-to-air missiles with multispectral seekers. At different aspects of the engagement, different sensors may have a lock on the target. Some combination of radar, acoustic, imaging, and IR sensors on the missile will be preferable to single-sensor missiles.

4. Missiles with long-range autonomous guidance

5. Missile warheads with increased blast radius

---

*These characteristics will apply to all interceptors, whether manned or unmanned. Teleoperated interceptors may be practical by 2010. If so, they would need capabilities along these lines.

6. Superior maneuverability. The best way to defeat stealth fighters may be through visual acquisition and guns (as in WWII).

7. Large numbers on combat air patrol at any one time. Since most detections of stealth aircraft and missiles will occur at short range, interceptors must be nearby to effect the intercept.

8. Light logistics. Air bases should be mobile. Squadrons must regularly redeploy to complicate enemy attack planning.

9. Stealth. Interceptors will need low signatures. High-signature interceptors won't survive in this environment. Both surface- and air-launched missiles (SAMs, AIMs) will increasingly be capable of autonomous tracking.

Air defense batteries will need these four characteristics:

1. Multispectral trackers and warhead seekers. At different aspects of the engagement, different sensors may have a lock on the target. Some combination of radar, acoustic, imaging, and IR sensors, both on the ground-based tracker and on the missile, will be preferable to single-sensor reliance.

2. Multiple shot and rapid reload. Assuming a low-signature target, individual SAMs may have low Pk. Advantage will accrue to systems capable of firing with only a marginal solution vice a system needing a high Pk shot (which might not happen). This will require inexpensive missiles and a rapid firing capability. Laser and/or directed energy weapons may prove to be weapons of choice.*

Note: This is a major issue. If cruise missiles are cheap, current missile defense concepts may find themselves on the adverse side of the expense ratio. For example, using an $800,000 Patriot missile to intercept a $100,000 cruise missile is grossly inefficient. This inefficiency will increase if more than one air defense missile is needed for each attacking cruise missile (due to a less than 1.0 Pk) and large numbers of air defense missiles are needed for every possible target area in case of saturation attack.

---

*Due to attenuation caused by the atmosphere (e.g., severe weather), a mix of weapon types may be preferable.

3. Mobility. Units must move daily to complicate enemy attack planning. Movements should result in minimal downtime with continuous positive C^{4}I.

4. Integrated effort with air defense fighters. Just as fighter interceptors shoot targets with friendlies in the engagement zone, air defense batteries must be able to fight in airspace occupied by friendly air defense fighters (joint engagement zone).

The overall approach should be to (1) fuse sensors to cue and track very low-observables, (2) integrate defense weapons so the most appropriate can be tasked whenever sufficient target information is available, and (3) wrap the entire aerospace defense system in an "OODA" loop of only a few minutes.

Degrade Enemy Cruise Missile Guidance

Classic counterair operations seek to destroy enemy aircraft either in the air or on the ground. Neither approach works well versus stealthy cruise missiles. Their inherent stealthiness makes radar and visual interceptions very difficult. Nor are attacks on cruise missile bases practical. Future peers will rely on mobile launchers; no readily identifiable bases will exist. For all intents and purposes, cruise missiles are less vulnerable to the two pillars of current defensive counterair doctrine: air interception and airfield attack. Cruise missile defense requires an additional approach.

The most exploitable weakness of cruise missiles may be their guidance systems. It may be possible to degrade enemy cruise missile effectiveness by targeting their navigation and terminal area guidance. For example, if attacking cruise missiles use GPS, defenders could manipulate the GPS unencrypted civil code within 1,000 NM of enemy launch areas. Because the GPS signal is very weak (.000001 watt), it is highly vulnerable to low-power jammers. Scattered 10- or 25-watt jammers could degrade GPS accuracy in specific areas. These jammers could either be on the ground or aboard HALE UAVs; this would complicate possible jamming countermeasures by varying the jamming direction. The GPS signal could also be manipulated by spoofing or by turning off the unencrypted civil code on those satellites within the field of view of route/target area (the US would retain military mode for US operations).*

Because of rapidly decreasing costs in inertial guidance (e.g., quartz or fiber-optic INS), these counters may have limited usefulness. The peer may not depend solely on GPS for guidance. The peer may also jam GPS theater-wide. However, if the peer uses external systems to either update or navigate cruise missiles, they'll present a possible weakness.

Terminal seekers "look" for specific patterns in the target area. These patterns may involve an infrared, radar, image, or acoustic signature (or some combination). Depending on the sophistication of the seeker's algorithm, it may be possible to spoof the terminal seeker. By understanding the pattern the seeker is programmed to find, decoy teams could devise returns which would attract the terminal seeker to benign areas. For this reason, understanding the seeker's algorithm should be a prime target of US intelligence. The job of spoofing terminal seekers should be a primary mission of air defense units (in addition to their physical interception mission).

Establish Ballistic Missile Defense

The principal advantages of ballistic missiles (speed, range, and mobility), make them integral to any weapons inventory. Assuming sensor-to-warhead target data transmission, a near-real-time (NRT) decision cycle, and warheads capable of identifying/tracking mobile targets (e.g., ships, TELs), ballistic systems offer unique and important military capabilities. Most importantly, they can kill targets with limited windows of vulnerability.

However, ballistic missiles have a major vulnerability. They offer a high signature. Ballistic missiles have a large infrared signature at launch and are radar reflective. They have minimal maneuverability. They can be tracked from launch through impact. Given these attributes, we can conceive of several ways to defeat ballistic missiles. Improved aerospace technologies

---

*As civil dependence on GPS increases, it will be politically impossible to "turn-off" GPS except in dire circumstances. War with a peer over a vital US interest, however, would meet this restriction.

(e.g., lasers, kinetic kill), integrated with improved computing technologies, offer considerable promise.

Key to ballistic missile defense (BMD) will be an integrated architecture which targets all aspects of a ballistic missile's life cycle. This life cycle includes: production; transportation; support personnel; C^4I; defenses; and the missile's three phases of flight (boost, post-boost, and terminal). This air defense architecture should have unitary command (except for point defenses) and be thoroughly exercised in peacetime. Finally, fixed, high-signature BMD will prove too vulnerable to the stealth cruise missile threat. Therefore, interceptors and sensors must also be mobile.

Unfortunately, a leak-proof BMD is probably impossible. As the Air Force Chief of Staff opined in 1995: "I'm not sure we're ever going to have 100 percent capacity to catch inbound missiles."[12] Because some leakage is probable, operations dependent on large force concentrations are untenable; we must devise military forces capable of dispersed operations.

Control and Exploit Space

Space will undoubtedly be a center of gravity in any future war with a peer. Space offers a medium for near instantaneous, cheap, worldwide communications. It offers the possibility of continuous surveillance of terrestrial events plus highly accurate positioning. These are war-deciding capabilities. If one side can exploit space for communications, collection, and positioning—while denying similar capabilities to its enemy—it will gain a decisive advantage. In 1986, the Chief of Naval Operations recognized this point directly:

> Today we know that in wartime, even in a conventional war of limited duration, the two superpowers would fight a battle of attrition in space until one side or other had wrested control. And the winner would then use the surviving space systems to decide the contests on land and sea. [13]

In general terms, war in space will mirror any other kind of war. It will have offensive and defensive aspects. Militaries will attack enemy satellites while trying to defend their own satellites. Space war will be fought over distances great and small. Targets will range from the surface of the earth (ground

41

stations) to GEO (geosynchronous earth orbit), plus every-
where in between. Weapons will be manned and unmanned,
kinetic-kill, and energy-kill. Environmental damage will
temper operations. Targets will include all facets of each space
weapon (e.g., C^2, infrastructure, production base, personnel,
and defenses). While the physics of space will dictate unique
weapons technologies, future war in space will involve goals
similar to those applicable in terrestrial warfare.

If left unchallenged, space architectures will provide
war-winning information. Prior to war, space sensors will
unobtrusively observe enemy force deployments, national and
military infrastructures, and physical characteristics of
potential areas of operation. These capabilities parallel much
of what the US Army stresses in METT-T (mission, enemy,
tactics, terrain, time). Two of the five (enemy, terrain) are
observable from space; mission and tactics can be inferred
from satellite reconnaissance. Having this information prior to
a war has immense military value.

During the conflict, space will act as the "grid" on which
critical information architectures "hang." Satellites will surveil
enemy maneuvers, assess friendly forces, aid positioning, and
facilitate communications. These capabilities will support both
the offense and the defense. They will help guide targeting
decisions while alerting terrestrial units of possible enemy
attacks. These are critical warfighting capabilities. They'll
enable every facet of combat, combat support, and combat
service support.

Space will also serve as a transit medium. Ballistic missiles
and some sensors will transit space on suborbital trajectories.
Either side in a war may wish to attack those platforms while
still in space. For these reasons, neither side in a major war
can allow its opponent unchallenged use of space.

Challenges to satellites will fall into four areas: ground- and
space-based lasers; exoatmospheric EMP/MHD/HPM*; jamming; and kinetic kill. These systems will vary in effectiveness.
The first two (lasers and frequency weapons) pose the lesser
challenge. Lasers can be negated through shielding. Proven

---

*EMP: electromagnetic pulse (<1MHZ); MHD: magneto-hydrodynamic (1-100MHZ);
HPM: high-powered microwave (100-200MHZ).

technologies can dissipate laser energy throughout the target. Frequency hardening (especially against EMP) is a well-understood, though expensive, process. Although a single type of shield cannot defeat all types of frequency weapons, shielding can protect satellites. To negate lasers and frequency weapons, military satellites should incorporate these features. For weight and cost reasons, however, COMSATs will not.

The latter two space interdiction threats (jamming and kinetic kill) are more problematic. Engineers can passively degrade the jamming threat to communications via frequency-hopping and narrowly-focused signals (e.g., EHF, laser). However, many forces will continue to rely on unfocused UHF signals. Such signals could remain susceptible to high-powered jamming. Their best defense may reside in using suppression forces against the jammers. Because jammers emit, they give away their precise location to antiradiation missiles. The inherently high signature of jammers is a substantial vulnerability.

Suppression of low-power jammers, however, will prove difficult. If the satellite is broadcasting or receiving low-powered signals, a low-powered jammer may suffice to interfere with its signal. This interference may render normal transmissions unreliable. Such low-powered jammers could be deployed in large numbers. They could also be mobile (space , air, land). Either technique would make suppression through interdiction difficult.* Also, jammer inventories will be important. Jammers produced and deployed by the thousands throughout the theater could overwhelm directional antijam filters installed on receivers.

The interdiction threat will vary depending upon target orbit. Because satellites in low earth orbit (LEO) can be reached by air-launched ASATs and space-based interceptors (e.g., Brilliant Pebbles), they will be vulnerable to frequent, short-warning, and relatively inexpensive attack. Their best chance for individual survival probably rests in increased

---

*A jammer aboard a satellite in close proximity to a GPS satellite would be exceptionally difficult to defeat. By co-orbiting the jamming satellite slightly ahead of the GPS satellite, an enemy would make some US interdiction efforts difficult. For example, a kinetic kill of the jamming satellite might scatter debris in the GPS satellite's orbit.

maneuverability and threat detection. By changing orbit when an ASAT/interceptor is en route, the target satellite may degrade the ASAT/interceptor's targeting solution (and its Pk). Unfortunately, this solution requires significant payload penalties. Extra fuel for maneuvering (and reacquiring the orbit) means less mission payload. Another possibility is co-orbital escort satellites. Just as fighters regularly escort reconnaissance aircraft and bombers within the atmosphere, escort satellites could escort/protect high-value collection satellites in space. The escort satellites would target ASAT/ interceptors attempting to intercept the protected satellite. These concepts of escort and maneuverability have significant downsides, however, principally in terms of magazine capacity and launch costs.

The best chance for architecture survivability will require a combination of maneuverability, escort, and—most impor- tantly—rapid replenishment.* In a sense, satellites in the next war should take a page from heavy bomber survival tactics in WWII. Although WWII bombers were readily detected and flew predictable routes (similar to satellites), their limited maneuverability and escorts provided enough protection for operations to continue. This protection sufficed because bombers were readily replaceable.** As soon as one bomber was lost, another took its place. For example, the Eighth Air Force lost, on average, 12 percent of its fleet each month in 1943 and 1944.[14] Despite these losses, Eighth AF heavy bomber inventories rose during that time. Although protection of the bombers was never very good, defensive measures and robust production sufficed to keep gains ahead of losses. LEO satellites in the next war should be as replaceable as bombers were in WWII.[15] Assuming that satellites in LEO will face similar attrition because they will operate in a similar situation (a high-value target, operating over enemy territory,

---

*Before dismissing this analogy out of hand, readers should put themselves in the places of military planners in 1925. From a 1925 perspective, the costs and attrition of WWII were horrendous. Nevertheless, they happened. The costs and attrition of a 2010 war with a peer will also be horrendous. Nevertheless, we must prepare with the scale of previous peer wars in mind.
**Bomber aircrew attrition, a most important factor in evaluating bomber operations, is omitted here because aircrew attrition has no direct comparison in satellite operations.

on a predictable route), a replacement regime on a par with that for WWII bombers is mandatory.

Satellite maneuverability would severely complicate ASAT targeting—if the enemy's space tracking capabilities were degraded prior to any maneuvering. Satellite locations are, at best, estimates based on studies of previous flight paths; change the flight path and you change the expected location. If the enemy is unable to construct a new flight path after a satellite's maneuver, it would be unable to predict an intercept location. For this reason, satellite tracking facilities will be prime targets in any war with a peer.

Satellites in GEO should experience higher survivability. They'll also be vulnerable to ASATs, but three constraints will mitigate these vulnerabilities. First, sophisticated ASATs (with extensive maneuverability and multiple sensors) will probably need a heavy booster to reach GEO. Heavy boosters require a considerable launch infrastructure. The limited number of such space-launch facilities could be targeted by nonlethal (e.g., conventional EMP) means.* Second, because easterly tracks along the equator are the most efficient for air launches into GEO, combat air patrols along likely launch tracks might degrade launch efforts. Third, an ASAT would need considerable time to climb to GEO for the intercept. During this time, countermeasures (e.g., maneuvering, a defense antisatellite weapon [D-ASAT], electronic intrusion) could occur. While satellites in GEO will be vulnerable to interception, these factors will make GEO satellites more survivable than satellites in LEO.

Offensive operations against enemy space systems will parallel these defensive measures. Enemy systems without the proper shielding, frequency management, maneuverability, and encryption will be vulnerable to interdiction. One caution: It is doubtful that the NCA will authorize a first strike on enemy space systems and infrastructure. We should assume that the enemy will strike first. Given this assumption, the US must have replacement satellites (and their launchers) ready at war's outset.

---

*Should nonlethal weapons fail, the NCA might approve a conventional attack on the ASAT launch facility.

Protecting the information flow will become as important as protecting the information collectors. Loss of either has the same effect. If US forces depend solely on satellites in GEO for data relay (e.g., SATCOMs, such as the military strategic and tactical relay satellite [MILSTAR]), it would present the enemy with a "single point failure" target set. By destroying the limited number of US military communications satellites, the enemy would make many US reconnaissance satellites less effective. To lessen this vulnerability, alternate information flows are needed. The most obvious is the civilian communication constellations projected for the near future. Iridium, Globalstar, and Teledesic, for example, promise to provide considerable bandwidth. However, these satellites will have two major weaknesses. They will orbit in LEO and won't be maneuverable. Enemy ASATs will take advantage of these weaknesses. They will likely "attrit" COMSATs in LEO (unless the enemy is also using the same LEO satellites for communications).

As with collection satellites, UAVs may offer an alternative.* A HALE UAV at 80,000 feet has a horizon of approximately 400 NM. Thus, it has line-of-sight connectivity with a similar UAV at the same altitude 800 NM away. A string of HALE UAVs, each 500–800 NM apart, could relay communications over several thousand miles. The last downlink could be to a ground station, connected via fiber optic cable with the national communications grid.** This downlink receiver could be either civilian or military, overt or covert.***

This UAV relay system would not *replace* a SATCOM system, but rather provide an *alternative channel* to satellites. It might also provide two additional benefits. First, a UAV architecture allows modifications of existing hardware on a daily basis. Unlike SATCOMs, which are "frozen" in R&D long before launch and, once launched, do not allow further hardware modifications, a UAV fleet could receive continuous hardware updates. Given the rapid pace of telecommunications

---

*HALE UAVs could also back up positioning (e.g., GPS) satellites.
**Any third wave country could supply the downlink station. All are connected with the information grid through fiber-optic cable.
***An ability to rapidly lay fiber-optic cable would further enhance this alternative.

technology, this is no small benefit. Second, an additional communications channel would induce a measure of stability in a crisis. As both sides postured for possible war, each would be tempted to preemptively attack the other's satellites. Given the vulnerability of satellites and their critical role in a future war, this "first strike" temptation may prove overwhelming. However, an alternative communications system would lessen the military advantage gained by a first strike on communications satellites. As a result, it could heighten crisis stability.

The concept of first strikes in space raises an important point. It is likely the US will operate with five disadvantages in any space war. First, the NCA will probably deny first strikes by US forces against enemy satellites, thus sacrificing the initiative in any war with a peer. Second, it is doubtful that the US will equip its satellites with nuclear power plants. If the enemy uses nuclear powered satellites, they will have decided power advantages over US satellites. Third, the enemy may have a faster acquisition cycle for satellites (dependent on solar and battery power). If it takes the US 10 years to design, build, and launch a satellite, and if the peer enemy can do the same job in five years, the US may be operating with inferior equipment. Fourth, a peer may aggressively weaponize space despite US and world opinion. This weaponization could include an extensive ASAT capability. Fifth, a peer may pursue an attrition campaign in space. A peer may build an architecture that is quite unlike the current US emphasis on expensive, multimission satellites. The peer might emphasize large numbers of single-mission, readily-replaceable satellites. Unit inefficiencies would be offset by their greater survivability in an attrition war. Should these five potential disadvantages prove true, the US disadvantage in any space war would be severe.

Integrate Intelligence, Surveillance, and Reconnaissance Systems

Aerospace collection and communications platforms come in three varieties: space-based; unmanned atmospheric; and manned atmospheric. In other words, satellites, UAVs, and aircraft. Sensors aboard these platforms also come in several varieties. Passive sensors include imagery intelligence (IMINT [photography, infrared]), SIGINT (exploiting an enemy's communications), and electronic intelligence (ELINT [pinpointing electronic emissions]). Active sensors include lidar and radar platforms (e.g., space-based wide-area surveillance [SBWAS], joint surveillance target attack radar system [J-STARS], AWACS, tactical reconnaissance aircraft [U-2, TR-1]). Each platform and sensor has unique capabilities and vulnerabilities. It makes considerable sense to integrate these systems into a whole.

Commanders need the capability to tap whatever intelligence, surveillance, and reconnaissance (ISR) sensor they deem necessary. If the peer should successfully target one aspect of the ISR system, other platforms must transparently assume that particular task. For example, both satellites and HALE UAVs are capable of wide-area surveillance and cueing. "National" systems such as DSP, "theater" systems such as J-STARS, and "tactical" systems such as an airborne early warning/ground environment integration segment (AEGIS) radar, also have overlapping capabilities. Should the enemy successfully interfere with one of these systems, commanders need the flexibility to task another system with replacement capabilities. The successor system would assume all or part of the mission. Such a transfer would require both a centralized tasking structure for ISR assets and a universal connection of all ISR assets to an overall C^{4}I system.

Another reason commanders need more than one ISR system is that they can use different types of sensors concurrently to decrease the effect of enemy spoofing. Enemy "targets" identified by electronic sensors might, when concurrently identified by imaging sensors, prove to be decoys. No sensor is ever perfect; but because independent probabilities are additive, two or more different sensors looking at the same target will give a higher-confidence product than a single sensor (or single type of sensor).*

AWACS, J-STARS, and TR-1 will have little utility in a war with a peer.[16] They emit continuous signatures, have little

---

*For example, suppose an ELINT satellite detects a target with 70 percent confidence. Concurrently, a SIGINT UAV identifies the same target with 50 percent confidence. Intelligence would assign an 85 percent confidence factor to that target (70% + [30% x 50%]) = 85%. This is the same method we use with missiles. One AIM-7 has a .7 Pk, two AIM-7s have a .91 Pk (.7 + [.3 x .7]) = .91.

maneuverability, and are highly reflective. They will be prime targets for stealthy interceptors. A peer's stealthy interceptors will likely penetrate into autonomous missile range of these aircraft.

In addition, a peer might field stealthy cruise missiles having an antiaircraft capability. These missiles would take cueing from land-based passive sensors (via triangulation), use antiradiation sensors for long-range tracking, then switch to a radar or IR sensor in the terminal phase.

This combination of attacking stealth aircraft and stealth cruise missiles would put high-value, emitting aircraft continuously on the defensive. They would contribute only intermittently to the overall campaign. High-signature aircraft such as AWACS, J-STARS, and TR-1 may still prove useful in rear-area defense roles, such as protecting an air base, port, stream of airlifters, or a convoy of ships.

Aerospace force structure should emphasize space-based systems and stealthy UAVs for 24-hour conflict surveillance, while de-emphasizing high-signature, high-value aircraft.* They should have redundancy between types of platforms, overlapping coverage among types of sensors, and connectivity with a common C^4I architecture. This integration will allow dominant battlefield awareness in a highly competitive environment.

Support the Information Campaign

Aerospace forces should expect heavy taskings in support of the Joint Force Information Component Commander's (JFICC) campaign. Satellites, UAVs, and manned aircraft will collect data on the enemy's information and C^2 architectures. Satellites and UAVs will relay this data to the JFICC's fusion and analysis centers. These centers will identify priorities and critical nodes within these architectures, which the JFICC will use to orchestrate offensive and defensive campaigns. In support of these campaigns, aerospace platforms (ASATs,

---

*To achieve continuous collection with a variety of sensors, the US should have hundreds of UAVs available for a war with a peer. In comparative terms, a few hundred UAVs would cost far less than the project J-STARS inventory. UAVs would also have significantly lower life-cycle costs.

missiles, bombers) will deliver munitions (both lethal and nonlethal) against JFICC-directed targets. Other military forces will also support the JFICC's campaign, but aerospace forces should expect sizable taskings.

This support will be a part of the theater CINC's normal apportionment process. The CINC will apportion a certain percentage of sorties to JFICC support (e.g., a certain percentage of UAV sorties on a certain day will fly in accordance with JFICC taskings). Just as aerospace forces are sometimes apportioned to support naval or ground campaigns, future information campaigns will see the joint force information component commander tasking aerospace forces in accordance with the theater CINC's overall guidance. The CINC will integrate this information campaign with ground, naval, and aerospace campaigns to effect a strategic victory.

At the same time, the peer enemy will be conducting its own IW campaign against the US. A prime target will be US military forces. Therefore, US aerospace forces must operate efficiently while under information attack.

The peer will undoubtedly attempt to corrupt information vital to US aerospace operations. The enemy's IW effort will probably center on four general areas: (1) deployment (e.g., the Federal Aviation Administration [FAA] network); (2) employment (e.g., the air tasking order [ATO], battle management); (3) surveillance (e.g., downlinks from ELINT satellites); and (4) logistics (e.g., supply requests). To mitigate the effects of such intrusion, aerospace forces must incorporate a series of defensive measures. These measures should include regular exercises in a corrupted information environment, software protocols which flag nonstandard inputs, redundant information links which check message fidelity while providing back up information routing, and extensive encryption that is changed regularly. Despite these efforts, we should expect at least modest success by enemy IW. We must learn to live with it—successful IW will be a given in future war. Just as army units have long operated under the threat of air attack, aerospace units must have the ability to operate while under information attack.

Key to successful operations in any war will be decision cycles. Both sides in a peer conflict will attempt to detect and task in near real time. Each will attempt to make snap decisions—faster and better than its opponent. Whoever builds the tighter decision loop will gain a significant advantage. This struggle for tighter decision loops will occur at all levels of war. Opposing fighter pilots (tactical), JFACCs (operational), and NCAs (strategic)—all will try to observe-orient-decide-act faster than their opponents. Each side will strive towards near-real-time decision cycles because they confer war-fighting advantages.

The advantage of near-real-time decisions carries with it a risk. Near-real-time decision cycles will require extensive use of automation and threat/opportunity triggers. By understanding either the algorithms inherent in the enemy's automated decision architecture or the key factors which trigger certain reactions, the commander can manipulate enemy responses. Therefore, a concerted effort to understand and exploit the enemy's decision process is mandatory. If effective, such an operation would initially drive the enemy toward bad decisions. After a series of bad decisions, the enemy would be forced to insert added cross-checks into its decision process, thus slowing down its decision cycle. As a result, snap decisions may be poor decisions if your opponent properly understands your decision process.

Conduct Offensive Strikes within the Enemy Homeland

In a future war with a peer, strikes on the enemy homeland are mandatory. The peer will have key facilities within its homeland integral to its war effort. These targets could include political and military leadership, weapons of mass destruction, command posts for operational forces, sources of national wealth, military sustainment depots, satellite ground stations, satellite tracking facilities, power projection forces (e.g., missile/bomber bases), and a national information network, among others. Successful strikes on these targets will have a critical effect at the strategic and operational levels of war.

Despite the critical nature of these targets, aerospace planners should expect significant *political restrictions* on these attacks. These restrictions will derive from a fear of nuclear retaliation. By definition, any peer will have

51

nuclear-armed ICBMs. A peer will probably threaten to answer any strike on its homeland, nuclear or conventional, with a nuclear-armed ICBM strike on CONUS. This threat may inhibit the NCA from authorizing a strategic air campaign on the peer's territory. At the very least, the NCA will want military options that don't include massive attacks on a nuclear-armed enemy's homeland.

Planners must reconcile the need to disable strategic targets within the enemy's homeland with probable NCA restrictions on doing so. A permissible option may involve precision strikes with *nonlethal weapons.* Although usually discussed in terms of low-intensity conflict, nonlethal weapons may have considerable utility in a war with a peer. Their employment against soft targets (e.g., electric grid, political organs, nuclear facilities, air defense C^4I nodes, fuel storage) could cripple the enemy's war-making capacity without presenting an excuse for a nuclear response.* For example, MHD bursts over stealth aircraft bases might cripple that fleet (at best) or its sortie generation (at worst). Spraying anti-fuel microbes on enemy air bases would also degrade sortie generation. Such strikes, if successful, would impair the peer's ability to orchestrate either strategic defenses or operational attacks.

Similarly, conventional EMP bursts near key electrical and information facilities might impair national C^2. EMP bursts near space launch facilities might deny the enemy access to space. In addition, some enemy targets, such as cruise missile facilities or military leadership, may prove impractical. They may be hardened or so dispersed as to be unreachable. In such cases, attacks will center on supporting infrastructures. Targets will include communications links and critical support (e.g., electricity). These nonlethal warheads could be delivered by stealthy cruise missiles launched from sea and air platforms during the day, or by long-range stealth bombers launched from CONUS at night. [17]

Bomber penetration of the enemy's homeland will require defense suppression and escort. After all, enemy defenses

---

*However, there should be no attacks with any type of warhead on any aspect of enemy ICBM forces (to include early warning satellites). Planners must avoid putting enemy strategic nuclear weapons in a "use-or-lose" situation.

against bomber attacks will have three factors in their favor. First, the enemy will know likely target areas (e.g., air defense headquarters within the capital city). This knowledge will narrow the focus of defense efforts. Second, peers may obtain limited cueing of bomber locations from long-wavelength radars or SIGINT intercepts. Partial information might suffice to enable visual interceptions. Third, high unit costs will keep stealthy bomber inventories low; loss of even one stealthy bomber will be significant. For these three reasons, bomber attacks on an enemy homeland warrant a pair of precautions. First, stealthy cruise missiles should precede bombers into defended target areas. Similar to the old Tacit Rainbow program, these missiles would autonomously suppress active defenses. They could also act as decoys, emitting signatures similar to those of bombers. Second, stealthy fighters should escort stealth bombers. The bomber needs some protection in case an enemy fighter acquires it. Because interceptions of stealthy bombers are a possibility, bombers need suppression and escort. Both the escort fighter and the defense suppression UAV must be stealthy to avoid compromising the bomber. Bombers should not penetrate alone.

Attack Enemy Invasion/Occupation Forces

War with a peer will probably involve contested territory outside the peer's borders. Unless someone figures out a way to occupy territory without putting soldiers on the ground, the peer's invasion/occupation forces will require large land forces. Such forces need equipment to take territory. They need tanks, artillery, and helicopters. These weapons need supporting trucks, ships, and logistics bases. All in all, invasion and occupation is a large-signature operation.

This weight is needed for a simple reason. To overcome modern defenses, massive numbers of mobile forces are a prerequisite. For example, Warsaw Pact plans for invading Western Europe envisioned massive numbers:

> It was estimated that in order to overcome the main line of defense, it was necessary to have at least a sixfold superiority

over the opponent. The breach of subsequent defense lines required only a threefold superiority. [18]

Unfortunately for the invading soldiers, massive numbers equate to a high signature—which results in a high chance of detection and targeting. This situation gives the defense an advantage.

Whether the invasion force moves by land or sea, the result should be the same. US space-based and atmospheric unmanned platforms will detect large-signature forces. Satellites and UAVs will provide awareness and cueing of operational-level enemy maneuvers. The primary sensor will be a phased-array radar with moving target indicator (MTI) capability aboard satellites in LEO. [19] MTI radars will search large areas and detect massed surface forces on the move. Electro-optical (EO) sensors aboard satellites in GEO will have sufficient resolution to keep a watch on main operating bases and probable avenues of attack. To complement satellite surveillance, UAVs would carry MTI/SAR (synthetic aperture radars) and EO sensors. UAVs would provide high revisit rates of specific areas of concern. UAVs would also replace satellite broad-area surveillance if the enemy degrades satellite operations. On land, possible attack avenues could be monitored by unattended ground sensors (UGS). These could be camouflaged and stealthily seeded by either aircraft or UAVs. Miniaturized ground sensors, incorporating robust microelectronics and communications, will sniff, watch, listen, and analyze. They could be densely distributed in high-interest areas and/or broadly seeded over areas of lesser interest.* After cueing by any or all of these sensors, UAVs would recce probable enemy formations and identify specific targets/locations for attack.

This data will be cross- or up-linked via satellite to a secure JFACC. JFACC will fuse this data (using wide-area automatic target recognition software) to rapidly identify enemy forces. Accompanying software would automatically assign priorities

---

*At sea, sound surveillance system (USN) (SOSUS)-type sensors would perform the broad area detection function. Probable detections would be interrogated by more precise sensors (UAV, satellite, J-STARS for surface targets; sonar buoys delivered by aircraft/UAVs for subsurface contacts).

to these targets based on threat and commander's intent.* After coordinating with the other component commanders (e.g., the land and naval component commanders), JFACC would then distribute taskings to worldwide units, which would conduct precise interdiction on enemy forces. After the attack, collection platforms would surveil the damage and, if necessary, reinitiate the process.**

Given these future capabilities, the theater CINC would probably levy the following operational goals on US aerospace forces:

• Degrade enemy invasion/occupation forces (e.g., stealth bomber strikes with brilliant munitions, cruise missile attack with an antiship warhead or mines).

• Hollow-out the enemy invasion/occupation force through attacks on logistics areas and LOCs (e.g., with cruise missile - delivered cluster bomb units [CBU], bomber-delivered mines).

Nonlethal weapons should prove especially effective against massed ground maneuver forces. Weapons such as high-power acoustic generators, high-power microwaves, EMP, anti-POL agents, and antirubber chemicals—applied against units in road march—should cause bunch-ups and disorganized advances. Conventional attack (e.g., a stealth bomber carrying 800 submunitions) could then inflict more permanent damage.

Satellites and UAVs will also identify large logistics bases. Once identified, JFACC will task the most appropriate munitions and delivery platforms to strike them. Logistics bases may prove an aggressor's greatest vulnerability. In the face of informed and precise attack, the enemy should be unable to develop the logistics infrastructure necessary for multidivision invasion/occupation.

US Army operations in Operation Desert Storm illustrate both these points. During the ground offensive, the US VII and

---

*This capability does not presently exist. A substantial development effort will be required to build this capability. This effort will take advantage of expected advances in parallel processing software and hardware, artificial intelligence, rule-based programming, novel database architectures, and networking.

**Bomb Damage Assessment (BDA) after long-range strikes will pose a severe challenge. The abilities to identify and strike targets at long range are only two-thirds of a war-fighting capability. The attacker must also know quickly and surely whether or not the attack succeeded. BDA ranges must equal strike ranges. One technique may be to lace explosives with signature chemicals visible to UAV and satellite sensors.

US Military Vehicles at Ad Damman, Saudi Arabia

XVIII corps required 1,600 truckloads of fuel and ammunition per day.[20] These supplies came from two logistics bases ("Charlie" and "Echo") which, themselves, took a month-long effort of continuous traffic to fill. When operating at full speed, an average of 18 trucks per minute arrived at these logistic bases.[21] Assuming an aggressor will have roughly similar logistics requirements, the signature and vulnerability of their logistics bases and convoys will make them highly vulnerable. Such massive logistics will be readily targeted in the new war-fighting environment. By hollowing out their logistics, US aero-space forces could immobilize enemy invasion/occupation forces.

An ability to retask strike missions en route would prove a great benefit in this environment. Because of the long flight times involved (e.g., some stealth bomber sorties will originate in CONUS), the tactical situation may change significantly between final aircrew briefing and time over target. When attacking maneuver forces, an ability to retask attackers en

Photo courtesy of US Army Transportation Museum

US Army Troopship, *General W. H. Gordon*, Departing Korea, 1951

route is mandatory. The type of information processing systems required to make retasking work will be the main difference between today's stealthy bombers (i.e., B-2) and those of 2010.

Targeting invasion/occupation forces is crucial to an overall approach to a future peer competitor. It is important for planners to recognize the advantages inherent in the defense in this 2010 war. Advances in information, C^2, penetration, and precision will make large surface forces highly vulnerable. This vulnerability will be highest during the initial stages of an invasion when the invader must mass to overcome indigenous defenses. It is at this stage that the US *must* engage the peer enemy. Failure to engage the enemy at this stage would prove disastrous. Once the enemy gains control of its objective, the US would find itself at a severe disadvantage. As the US tried to mass forces to capture the lost territory, its logistics , convoys, build-up areas, etc., would come under heavy attack. In essence,

Overhead View of Tent City at Sheikh Isa, Bahrain, During *Operation Desert Storm*

the process outlined in this paragraph would be turned against the US.*

Avoid Deployment of Critical Targets within Range of Enemy Stealth

It has always been a sound tenet of military doctrine to keep friendly forces—to the maximum extent possible—outside the range of lethal enemy systems. Units close with the enemy only when necessary to accomplish specific objectives. This tenet will not change in a future war with a peer. If anything, the concept will become even more important as information, C^2, penetration, and precision capabilities increase. Complicating this situation will be concurrent increases in weapon range.

---

*For this reason, naval platforms (arsenal ships, aircraft carriers, SSNs with cruise missiles) will have secondary roles to long-range, land-based bombers. The time needed to sail ships with their escorts to the AOR may exceed the vulnerability window of the peer's power projection force.

In a future war with a peer, prudence dictates that aerospac e forces base few assets within range of enemy stealth . To the maximum extent possible, US aerospace forces must base *outside* the range of enemy stealth systems. For example, if the peer enemy's cruise missiles have a 1,000-NM range, US aerospace forces should base >1,000 NM from the likely operating area(s) of these missiles. High-value airborne platforms (e.g., AWACS, J-STARS, Rivet Joint) would launch and recover from bases >1,000 NM from the enemy. The US goal should be to concentrate fire, not forces.

Such basing will be possible only if the US aerospace inventory emphasizes long-range operations. Aircraft will need long legs for flights from rear area bases to enemy targets, and aircrew ratios must support long sortie durations.* Inventories of aircraft and munitions must be sufficient to deliver effective, sustained firepower on the target set from bases >1,000 NM away. Deployment kits must support extended operations with minimal support. This will enable split unit operations and frequent changes of bases. Finally, because of enemy aero- space defenses, these long-range strike systems must be stealthy.

Having said that, however, we must understand that significant forces will still have to operate routinely within range of enemy deep-strike systems. For example, the CINC may deploy surface-to-surface missiles into the theater to threaten time-critical targets (TCTs). In addition, UAV, C^2, and BMD units will deploy close to the fight.

Several measures will increase their survivability. Their arrivals in theater should be covered by ground and airborne air defenders.** They would deploy/disperse/camouflage during darkness. Most importantly, forces deploying within range of enemy stealth must be mobile. They must constantly shift their location, if only by a few miles. If they simply deploy to one location and sit pat, enemy surveillance systems will eventually pinpoint their location; enemy deep-strike stealth will likely penetrate with precision. Immobile facilities necessary for

---

*Although aircraft may fly multiple sorties in one day, aircrews will not.
**Air defense fighters over airheads would attempt to visually acquire attacking enemy stealth fighters. We should assume stealthy cruise missiles will *not* be observable in flight. Individual missiles may be visually acquired and shot down, but this will be an exception, not the rule.

operations should adopt ship and tank antimissile defenses: (1) kinetic-kill (e.g., Phalanx) or directed-energy weapons (DEW) for point defense; (2) reactive armor to decrease warhead explosive effect; and (3) decoys.

Position JFACC in CONUS

Our current C^2 CONOPS deploys the JFACC* to the theater of operations. Forward deployment has the advantage of allowing face-to-face contact between the theater commander (the CINC) and the JFACC. It also fosters personal relationships with coalition partners. However, this deployment has two significant downsides.

First, deploying JFACC to the theater puts a high-value/high-signature target within range of enemy stealth systems. As the key aerospace battle manager, the JFACC will top the enemy's target list.** With its large infrastructure (e.g., antennas, tents, vans) and robust communications, sooner or later the enemy will pinpoint the JFACC's location.[22] If this location is within range of enemy stealth systems, those systems will eventually penetrate US defenses and precisely attack the JFACC's headquarters.

Second, JFACC is unable to direct the campaign while physically deploying to the theater. While en route, and until the key staff with its equipment and defenses are in place, JFACC will have neither the knowledge nor the connectivity to orchestrate an aerospace war. This delay is a critical shortcoming. The peer enemy is most vulnerable during its invasion phase. Logistics are massed; routes of march exposed. Giving the enemy a "breathing space" during its most vulnerable time is a questionable CONOPS.[23]

A solution to both problems is to base the JFACC in CONUS.*** This basing would keep the JFACC outside the range

---

*As stated previously, this book uses the term "JFACC " to encompass all aerospace C^2 above wing level (e.g., AOC, TACC, ROCC, LRR, ASOC).

**Just as the Iraqi Air Defense HQ was a high priority during the opening phase of Operation Desert Storm—which the US attacked with stealth systems (i.e., F-117).

***The logic for putting the JFACC in CONUS may also apply to other component commanders. For example, the naval component commander in the 1991 Gulf War owned forces operating in the Persian Gulf, the Arabian Sea, and the Red Sea. His presence aboard one ship operating in one of these locations did little to improve his decision making over other NAVCENT forces.

of enemy stealth systems and avoid creating a fixed, in-range, high-value target for the enemy. It would also allow immediate planning/tasking of the air campaign. There would be no delay imposed by waiting until JFACC (with its defenses) has deployed and set up operations. Instead, the JFACC could begin directing the air campaign immediately. Planners would have immediate access to all-source intelligence. A CONUS JFACC would allow well-exercised connectivity with combat units (e.g., fiber-optic cable connections with CONUS-based stealth bomber wings).[24] They could take advantage of CONUS databases and expertise; JFACC computers could be hardwired to a secure information net. All data relayed by satellite (including data from national systems) would downlink to a fixed JFACC facility. It would fuse the data, filter out extraneous material, and distribute distilled information. In essence, JFACC would immediately have the exercised expertise to turn information about the situation into knowledge about the war. After running computer simulations to determine the best tactical options, JFACC would issue the ATO. This centralized ATO would direct all air assets, whether based in-theater, in CONUS, or in adjacent theaters.

This approach is compatible with the current communications concept of "smart push, warrior pull." If JFACC were colocated with the worldwide intelligence manager, unit taskings and the applicable intelligence information could be distributed concurrently ("smart push"). Intelligence officers sitting alongside the operational tasking officers would "attach" the requisite intelligence information.* Issuing both the tasking *and* the accompanying intelligence would decrease ATO cycle times, as units could immediately begin mission planning based on the most current information (as opposed to drafting an information request and waiting for the response). It would also provide the most appropriate information to the units whether or not the units were aware of its existence. Finally, it would provide an alternative to the tendency to make "everything" available to the tactical level.

---

*Computers would handle most of this function. Certain types of targets would automatically generate certain types of intelligence. They might also automatically generate certain intelligence taskings.

61

The tendency to make everything available to the warrior has the potential for overloading users and transmission means. Of course, units would retain the authority to query the database for additional specific information ("warrior pull") as they saw fit.

Complexity is another factor arguing for a CONUS JFACC. Orchestrating an aerospace war is anything but simple; it is extremely complex. Weapons are air-, space-, land-, and sea-launched. Targets are fixed and mobile, hard and soft, terrestrial and space, strategic and operational. Some platforms move at tens of thousands of miles an hour (in space); others move at a few knots (at sea). Squadrons are scattered around the globe, their strike packages coming from equally scattered units. Support comes from an alphabet soup of agencies: CIA, DIA, CIO, NRO, DISA, DMA, NSA. Data requirements are measured in terabits. If JFACC must deploy to the theater, this orchestration must be accomplished by a mobile C^4I structure— adding another factor of complexity to an already incredibly difficult process.* Establishing a permanent CONUS JFACC would delete at least this additional level of complexity.**

Airlift Critical Supplies and Spare Parts into the Combat Area

The CINC will probably direct a substantial airlift flow into the combat theater to support its accompanying C^4I, component forces, and indigenous forces. Airlift operations must reconcile their CONOPS with the peer's information, C^2, penetration, and precision capabilities. As an entering assumption, airlift planners must allow for the probability that all large airlift operations will operate under some measure of enemy observation. As a result, airlifters operating within

---

*Even if JFACC is already positioned in theater (e.g., the HTACC at Osan Air Base, Republic of Korea), a back-up facility must be capable of assuming this complex orchestration. However, any HTACC back-up will not be as effective; it's unreasonable to expect equal capabilities from back-up facilities/personnel. Thus, putting JFACC at Osan provides the DPRK a high-value target.

**Another argument could be standardization. Given rapid advances in information technologies, it is likely that theater commands (EUCOM, PACOM, CENTCOM) will build different C^4I structures. This will hamper training. Units will have to prepare to interact with several different command structures.

range of enemy weapon systems must also operate within the enemy's OODA loop. They must be able to arrive and depart before enemy C⁴I can detect the airlifters, direct an attack, and deliver warheads onto the target. By operating within the enemy's OODA loop, airlift sorties can flow into the theater.

This will require minimal ground times by all sorties into bases within range of enemy systems.* Rapid off-loads are mandatory in this environment. Arriving forces would disperse immediately after landing.

Civilian airlifters (CRAF) will have little use in this environment. They are neither configured for rapid off-loads ("roll-off") nor hardened against EMP. Their need for long ground times to off-load cargo will place them outside the enemy's OODA loop. The enemy will have time to detect, launch, and strike civilian cargo transports on the ground. If these strikes carry conventional EMP warheads, precision will not be necessary. An EMP blast within a mile of a civilian airplane, with its unshielded fly-by-wire controls, could disable that airplane. Despite the heavy costs incurred by relying exclusively on military airlifters, they're required for airlift operations during a war with a peer. In addition, because any military airlift fleet will have a finite size, airlift requirements for aerospace forces must fit within a much smaller ton-miles/day capacity than is presently assumed.

Summary

This chapter has discussed operational concepts for US aerospace forces in a future war with a peer around the year 2010. In such a war, both sides will undoubtedly possess thousands of state-of-the-art aerospace weapons. These weapons will include stealth systems (cruise missiles and manned fighter-bombers), information systems (surveillance and communications), nuclear weapons, and ballistic missiles with intercontinental range. A peer enemy will also possess

---

*If the threat is enemy cruise missiles, ground times could be as much as an hour (due to time of flight). If the threat is from electromagnetic launchers (railguns), ground times should be less than 10 minutes.

sustainable and redundant military capabilities. Because of the geopolitical environment, it is safe to assume the majority of conflict will occur on the enemy's borders and that these borders will be several thousand miles from the CONUS.

This future war will be fundamentally different from those possible today. The biggest difference will lie in the inability of aerospace defenses to protect high-signature forces from attack. In a future war with a peer, we must assume stealthy cruise missiles and aircraft of both sides will penetrate aerospace defenses in significant numbers. These systems will target critical vulnerabilities (due to modern surveillance systems) and will hit what they target (due to modern precision). Each side will also have near-real-time C^2, redundant capabilities, and long-term sustainability. Unlike Operation Desert Storm, critical nodes in these systems will operate from sanctuary; the threat of nuclear retaliation will place restrictions on homeland strikes. Taken in aggregate, this environment differs markedly from current conditions. It will require fundamental changes in our concepts of operation. In a sense, our situation is similar to the one faced by military strategists during the interwar period.

Between WWI and WWII, developments in aircraft and armored vehicles fundamentally changed the conduct of war. Those who succeeded in opening stages of that war were the ones who adjusted their CONOPS to fit the new technological environment. If we posit a major war 15 years from now (in 2010), we should expect similar magnitudes of change. Driven by stealth and information technologies, the magnitude of difference between a war today versus one in 2010 could be comparable to the difference experienced in the interwar period (1925–1940). The time interval is the same. If the WWII analogy holds, critical weapons and CONOPS, proven in the past and relied on today, will become obsolete over the next 15 years.

As we project a war in this environment, two themes keep repeating. The first theme is that we must engage peer aggressors when they are in the invasion mode. This is when they are most vulnerable. The new generations of weapons can detect and destroy massed surface forces on the move. If we fail to engage the aggressor immediately, we'll find ourselves on the adverse side of this exchange ratio; we'll be in the invasion mode, trying to move large forces in the face of advanced enemy information, C^2, penetration, and precision systems.

The second theme involves the question of a CONUS-based versus theater-based JFACC. This question requires extensive examination. Our preliminary judgment leans towards the former because forward deployed headquarters are vulnerable, require time to set up, and have inherently poor connectivity (compared to a centralized approach). Spending valuable hours and sorties to move a headquarters—especially one that will have inferior communications—within range of the enemy's missiles is a questionable way to operate. Furthermore, the theater will extend over millions of square miles. Critical assets, such as satellite control and long-range bombers, will base outside the theater. There is little to be gained by placing JFACC within several hundred miles of some subordinate units; practically all communication between them will "bounce" off satellites. Theater-to-CONUS communications will take the same time and routing as theater-to-theater. The important question is, where can the JFACC get the best information? A centralized, CONUS-based JFACC structure seems the best alternative.

The following aspects of future peer warfare deserve special emphasis:

• Aerospace defenses must anticipate massive numbers of low-signature attackers. If unit costs of cruise missiles decrease to the $100,000 range, both sides will likely employ large numbers of very low-observable attackers.

• Offensive aerospace forces must de-emphasize high- signature, theater-based assets. Their attrition in the emerging environment will be sufficiently high to preclude high-tempo operations.

• Planners must take steps to induce greater crisis stability into the US force structure and CONOPS, especially with regard to space. Absent greater redundancy and more effective defenses, the US could find itself in a first strike predicament during a crisis.

65

• Planners must avoid deployments of critical fixed targets within range of enemy stealth. Fixed facilities will face an unacceptable risk of destruction by precision stealth systems.

• Planners should integrate satellites and UAVs for communications, navigation, and surveillance. UAVs promise sufficient loiter times—and survivability—to accomplish these missions. Integration will allow rapid substitution and reduce the effects of deception (through cross-checking).

• Space will be a center of gravity in any future war with a peer. Both sides will rely on satellites for communication, positioning, and collection. Satellites in LEO will be particularly vulnerable. They will require both active and passive defenses, including shielding, maneuverability, rapid replacement, frequency management, and redundancy. Satellites in GEO will be secure if enemy launch facilities capable of GEO reach are destroyed.

• Future peer aerospace forces will include stealthy interceptors. As a result, high-signature atmospheric platforms (e.g., AWACS, J-STARS) will not thrive in a future war with a peer.

• JFACC should base in CONUS. Fixed, permanent basing will allow immediate tasking of worldwide assets while excluding a high-value, high-signature target (JFACC HQ) from the range of enemy stealth systems.

• JFACC will provide NRT information on allied and enemy maneuvers to allied forces. This transfer will require specialized equipment and liaison teams.

• Satellites are lucrative targets absent (active and passive) defensive measures.

• Defensive counterair must emphasize sensor fusion. Because a significant portion of the enemy aerospace force will be stealthy—and stealth systems in flight will intermittently reflect and emit—a thoroughly fused sensor network is important. It holds the possibility of successful detection and targeting. This system must be mobile to preclude targeting by ballistic missiles.

• Degrading enemy cruise missile guidance will be a top priority. By manipulating external guidance systems (such as GPS), and by positioning decoys in the target area, defenders will attempt to exploit any algorithm weaknesses in the enemy system.

• Planners must devise a concentrated offensive against key targets within the enemy homeland. C^4I is the highest priority. The NCA will probably restrict these strikes due to the threat of nuclear retaliation. For this reason, nonlethal weapons, delivered by stealthy bombers and cruise missiles, will assume a leading role.

• The aerospace campaign will attempt to deny enemy invasion/occupation, primarily through long-range bombers with precision munitions and cruise missiles. Logistics will be the most lucrative target set.

• When airlifting critical supplies and spare parts into the combat area, operators must minimize ground times. Depending upon distance from enemy missile launchers, ground times will usually be measured in terms of minutes, not hours. Airlift must be capable of efficient operations despite an information-corrupted environment (to include nonavailability of GPS).

• Future aerospace forces will attack critical enemy targets in a parallel fashion, denying their ability to adapt or repair in advance of subsequent strikes. Their goal will not be attrition; they will attempt to paralyze enemy C^2.

• Defenses will take advantage of ballistic missile vulnerabilities (large infrared signature at launch, radar reflective in flight, minimal maneuverability). Having said that, a 100 percent shield is probably impossible.

• Aerospace forces should expect heavy taskings in support of the joint force information component commander's (JFICC) campaign. Operations will center on (1) destroying nodes (such as collection platforms, relay networks, and fusion centers) and (2) distorting information by viral insertion and spoofing. Because the enemy will also conduct IW, aerospace forces must prepare to fight in an information-corrupted environment.

These themes should guide aerospace planning for a future war with a peer. Because evolving technologies will allow thousands of precision strikes per day, planners must devise a new CONOPS to take full advantage of this new capability.

Notes

1. *The Military Balance, 1994–199 5* (London: The International Institute of Strategic Studies, Brassey's, 1994). The US budgeted $261.7B for defense for FY 1994. The next eight largest defense budgets (in order): Russia ($79B); Japan ($42B); France ($35B); United Kingdom ($34B); Germany ($28B); Italy ($16B); South Korea ($14B); and Saudi Arabia ($14). Total: $262B. China's military budget is difficult to state with precision. *The Military Balance* estimates somewhere between $7B to $27B (see page 170). *Note:* Dollars for defense are not an absolute gauge of military capability. They are only a rough indicator. However, ratios of four, eight, or 20 to one suffice to preclude military equivalence.

2. Peter M. Senge, *The Fifth Dimensio n* (New York: Doubleday, 1990), 313–48.

3. Richard G. Davis, *Carl A. Spaatz and the Air War in Europe* (Washington, D. C.: Center for Air Force History, 1993), 358–60. In January 1944, Eighth AF's escort fighters changed their CONOPS from "close escort" with bombers to "ultimate pursuit."

4. Brilliant sensors can discriminate between targets (e.g., identify a tank versus a truck).

5. Full ATR under all weather conditions is foreseen within 10 years. See *Aviation Week & Space Technology*, 6 February 1995, 20.

6. "USAF Almanac 1995," *Air Force Magazine*, May 1995, 50. Figure is TAI (Total Aircraft Inventory).

7. Office of the Secretary of the Air Force, *Gulf War Air Power Survey*, vol. 5, 1993. (Secret) Information extracted (Table 75) is unclassified.

8. Proprietary conversation between OSD/NA and a corporate vice president of a major US defense contractor, May 1995.

9. Air Force Manual (AFM) 1-1, *Basic Aerospace Doctrine of the United States Air Force*, vol. 1, fig. 2-2. "Centralized Control/Decentralized Execution" is a "Tenet of Aerospace Power." "Execution of aerospace missions should be decentralized to achieve effective spans of control, responsiveness, and tactical flexibility."

10. AFM 1-1, vol. 2, defines air supremacy as "That degree of air superiority wherein the opposing air force is incapable of effective interference." Air superiority is "That degree of dominance in the airbattle of one force over another which permits the conduct of operations by the former and its related land, sea, and air forces at a given time and place without prohibitive interference by the opposing force."

11. During the 1991 Gulf War, the USAF deployed prototypes of the E-8 J-STARS to the Gulf with six software specialists. These specialists wrote 17 upgrades to the J-STARS software in order to adjust expected conditions to the realities of the Gulf. Many viewed this deployment of code writers with a weapons system as a unique event, necessitated by the rush to employ J-STARS ahead of schedule. However, future deployments of software experts with weapon systems may become the norm. Alternatively, the code writers could remain in CONUS if they are well connected with the deployed systems.

12. USAF Chief of Staff Gen Ronald R. Fogleman, quoted in *Inside the Air Force*, 3 February 1995, 9.

13. Colin S. Gray, quoting Adm Carlisle A. H. Trost, USN Retired, "Space Power Survivability," *Airpower Journal* 7, no. 4 (Winter 1993): 27–42.

14. Davis, appendixes 7 and 8.

15. Message, 122007ZJUN95, Chief of Naval Operations Weekly Update, 05-25. A rapid replacement CONOPS would have to overcome two current cultural impediments: (1) our reliance on only two launch centers and (2) the considerable time we take to check out vehicles prior to launch and satellites after reaching orbit. For example, the fifth UHF Follow-on Advanced Communications Satellite, launched 31 May 95, needed two months of on-orbit testing before being turned over to Naval Space Command for operational service.

16. On 25 April 1995, USAF Chief of Staff Gen Ronald Fogleman stated: "I see a future without AWACS. Instead, space-based assets will be providing the air picture and (have the benefit of) not tying up tankers." Quoted by Tanya Bielski, "Air Force Chief Embraces Information Warfare," *Defense Daily*, 26 April 1995, 125.

17. For a broad discussion of nonlethal weapons in strategic attack, see Jonathan W. Klaaren and Ronald S. Mitchell, "Nonlethal Technology and Airpower: A Winning Combination for Strategic Paralysis," *Airpower Journal* 9 (Special Edition, 1995): 42–51.

18. Interview with Gen P. S. Grachev, Russian Minister of Defense, *Izvestia*, 2 June 1992, 2. Quoted by Michael M. Boll, "By Blood, Not Ballots: German Unification, Communist Style," *Parameters*, Spring 1994, 66.

19. William P. Delaney, "Winning Future Conflicts" (Unpublished manuscript, MIT Lincoln Laboratory, 15 October 1992). Essentially a space-based version of J-STARS. Three-meter resolution would suffice to identify military convoys. An eight-satellite constellation would give 30-minute revisit times for areas at 50° latitudes. Each satellite's returns would be fused to enable constant dwell time.

20. William G. Pagonis and Jeffrey L. Cruikshank, *Moving Mountains* (Boston: Harvard Business School Press, 1992), 147. Also, even if current efforts to reduce logistics requirements succeed, logistics signatures will remain high. For example, if these two corps reduced fuel and ammunition requirements by 25 percent, they would still expose a target set of 1,200 trucks.

21. Ibid., 146.

22. During Operation Desert Storm, the JFACC headquarters in Riyadh consisted of 2,000 personnel. See *Gulf War Airpower Survey*, chap. 5.

23. If the war's in an immature theater, this "breathing space" could be quite long. "The several months needed to create the *ad hoc* communications, tasking, processing, and information reporting systems used in Desert Storm represents an unacceptable readiness posture." Report of the Defense Science Board Task Force on Global Surveillance, December 1993, 3–8. (Secret) Information extracted is unclassified.

24. According to Lt Gen Carl O'Berry, Headquarters AF/SC, a high-bandwidth fiber-optic network connecting all 111 AF bases in the CONUS would cost $1.3B. See "Bandwidth or B-2 Bombers?" in *Government Computer News*, 5 June 95, 68.

Chapter 3

Niche Competitor

Paul Bracken describes the United States as experiencing an "Indian summer in national security."[1] For the first time in half a century, America faces few threats to its vital interests. Its enemies are weak while its friends are strong. America's borders are secure, its economy is growing, and its military is far and away the finest in the world. All in all, America is in a relatively comfortable situation.

This is not to say threats don't exist. Given certain circumstances, small states such as North Korea or Iraq could threaten US interests. But in relative terms, such threats are manageable. During this Indian summer in national security, all threats to vital American interests can be managed by current US security mechanisms.

But as Bracken's metaphor implies, this condition won't last. Just as winter inevitably follows Indian summer, more substantial threats to America's vital interests will eventually arise. As the previous chapter outlined, these threats could take the form of a peer competitor. Fortunately, for the reasons stated, a peer competitor is unlikely over the next 15 years. A more likely threat within the next 10–20 years is a niche competitor.

A Niche competitor is a state (or alliance) that combines limited numbers of emerging weapons with a robust inventory of current weapons, then develops an innovative CONOPS to best employ this mix. Examples of possible niche competitors include Iraq and North Korea.

There are five key points to remember when envisioning a niche competitor.

First, a niche would always be militarily inferior to the US. It would have a weaker military *and* it would have a weaker strategic position. By the former, we mean the niche would never have the breadth and depth of weapons available to the US. A niche could never hope to slug it out toe-to-toe with the US. It would inevitably lose an all-out war. Its goal would be to raise the cost of US involvement beyond an acceptable level. A niche would seek to effectively challenge US interests in its region by making the US military response sufficiently costly to either deter initial involvement or dissuade further involvement.

By the latter point, we mean a niche would find it difficult to close out US options; that is, to decisively knock the US out of a war. Because of its strategic depth and wealth, the US will always have the option of "revisiting the decision." The US could lose the initial campaign and withdraw, then return to the fight after rearming and restructuring. A niche, on the other hand, wouldn't have these options. Once it loses to the US, the niche would find it impossible to mount another campaign of equal or greater intensity.

Instead of counting on an absolute defeat of the US, a niche competitor's best course would be to encourage the US to avoid the fight. The US will probably do so when it perceives its possible costs exceeding its gains. The prospect of high casualties or a drawn-out conflict may affect this perception. Such an election is different from being *forced* out. Conversely, the niche will present multiple strategic centers of gravity to US attack. Its leadership, industrial base, national infrastructure, population, and key military forces will be reachable by US aerospace forces from the first day of the war onward. It will be far easier for the US to close out a niche state than for the niche to decisively defeat the US.

Second, the niche will present operational centers of gravity to attack. We can assume the niche is doing something *outside its borders* that is contrary to substantial US interests. That is the *casus belli* for US military involvement. The invasion/occupation involved in this aggression must be of sufficient size to gain and hold territory.* The invading forces would require personnel and equipment numbering in the tens of thousands. These operational forces would present numerous critical targets for attack. Their detection and targeting would be a prime mission for US aerospace forces.

Third, many nations have the capacity to attain niche status. Unlike a peer competitor, a niche seeks to develop a

---

*Examples could include second invasions of Kuwait by Iraq or of South Korea by North Korea. We're *not* talking about civil wars against an insurgent or nonstate enemy.

proficiency in only a few mission areas, as opposed to many. For example, a niche may invest only in stealthy cruise missiles, space interdiction systems, and data fusion centers, while eschewing competition in sea control, air superiority, R&D, airlift, and amphibious operations.

Because its military capabilities are considerably less, a niche can get by on a smaller resource base. Smaller resource requirements mean there are literally dozens of nations capable of fielding a niche-competitor military. Examples of possible niche competitors include the two countries used in current MRC planning (Iraq and North Korea), plus Australia, Brazil, Chile, India, Iran, Nigeria, Pakistan, South Africa, and the Ukraine, among others.* In essence, any nation with the wealth to buy limited numbers of emerging weapon systems is a potential niche competitor. While a niche would not have the size or sophistication of the US military, it might be able to frustrate limited US objectives in its region.

Fourth, a niche would have to be capable of doing more than fielding state-of-the-art weapon systems. Modern weapons underwrite the ability to compete in the new warfare environment, but are not enough in themselves. To take full advantage of the capabilities inherent in emerging weapons, a niche military must be able to adjust its CONOPS as well as its inventory. For example, a niche must do more than simply buy information weapons. Rather, it must integrate information war systems with the rest of its inventory in a synergistic way. We must expect the niche to employ and control new technologies in innovative ways. These ways might differ markedly from their past doctrine.

Finally, unlike war with a peer, war with a niche will be a "come as you are" war. The absence of risk to US vital interests would preclude domestic American support for a rapid buildup. The prospect of war with a niche would probably have little effect on the future years defense plan (FYDP). In addition, warning time would be shorter than for war with a peer. The niche would need little time to field its limited number of emerging weapons. These two factors mean

---

*These nations are listed solely for illustrative purposes. There is no intent to imply conflict between one of these nations and the United States.

the US would likely have only its existing units at its disposal. Some acceleration of procurement immediately before hostilities is possible, but a rapid US rearmament is unlikely. US employments would largely mirror those planned and exercised in peacetime.

Niche competitors will face a similar situation. They'll have military requirements unrelated to a war with the US. For domestic and regional reasons, niche competitors will be unable to focus their military efforts solely on defeating the US. In fact, only a small proportion of their military will be optimized for defeating US forces. Niche states will have more important military missions than just war with the US.

Take present-day Iraq as an example. Iraq uses its military for two missions which have nothing to do with the US. Its military's highest-priority mission is internal security, which is Saddam's number one security problem. Iraq's military manpower, resources, and CONOPS must support this mission. Officers are promoted based on loyalty and their ability to effect internal security. The second-priority mission for Iraq's military is security against regional foes. Turkey, Syria, and Iran all pose credible threats to Iraq. Iraq's military must maintain the capability to deter those countries. For both these military missions (internal security and regional defense), manpower-intensive, attrition militaries are effective. Tanks, Scuds, infantry divisions, helicopters, and nonstealthy fighters are suited to these missions.

Fighting the US is a third-priority mission for the Iraqis. Unfortunately for them, however, the military forces and CONOPS required to defeat the US are ill-suited to the first two missions. But the fact that Saddam Hussein's forces are ill-suited to fight the US doesn't mean he can afford to dispense with them. Quite the contrary. Saddam, like other potential niche competitors, must maintain internal security, defend his borders against larger neighbors, and remain ready for war with the US. That's a difficult charge—and a very expensive one. Iraq's best solution is to lay a limited number of emerging systems on top of an existing force structure that was built with two higher priority missions in mind. Iraq *cannot* build a military focused solely on defeating the US. Other niche competitors would face a similar situation.

In summary, a niche could compete with the US by employing bits and pieces of advanced technology along with a robust inventory of traditional weapons. It would integrate these weapons using innovative strategies to offset the greater military breadth and depth of the US. Its goal would be to persuade the US to leave the conflict (as opposed to seeking a decisive military victory). The niche would exploit asymmetries in strategic culture, geography, and political/military objectives. Warning time for this war would be much shorter than that envisioned for a peer conflict.

Environment

When projecting a future conflict with a niche competitor, the United States must expect the enemy to field a mix of emerging and previous systems, as well as to use an asymmetric method of employment. The types of information, C^2, penetration, and precision systems, as well as the number and size of their weapons of mass destruction help to distinguish niche competitors. Table 3 compares the capabilities of a peer competitor to those of a niche competitor.

Table 3
Niche Competitor Compared to Peer Competitor

	PEER COMPETITOR	NICHE COMPETITOR
Information	Indigenous, Dedicated	Third Country, Commercial
C^2	NRT, Redundant, Automated	Delayed, Nodal, Hierarchical
Penetration	Multisystem	Single System
Precision	Autonomous Guide (e.g., Terminal)	External Guidance (e.g., GPS)
WMD	Hundreds. Can Reach USA.	<10, Theater Reach
Size	Large, Strategic Depth	Small, Little Depth

Emerging Systems

In general terms, a niche might have emerging systems with access to commercial satellite (COMSAT) networks (communications and surveillance), modern C^2 systems, stealthy cruise missiles (equipped with either warheads or sensors), advanced missile guidance, and advanced conventional munitions. In more specific terms, a niche's emerging systems would emphasize information, C^2, penetration, and precision.

A niche enemy will use a mix of civilian and military *information systems* for military purposes. It will use civilian surveillance satellites to detect large US force movements. Data obtained from civilian sensors will not be near real time (NRT); it may be several days old. Despite its age, such data will prove useful in identifying large, fixed, build-up areas (e.g., airfield parking ramps, logistics points, lines of communication, ports).* Civilian communication satellites will relay military data and instructions. Cruise missiles will have surveillance and communications packages to augment satellite coverage.

Owing to expected advances in the civil sector, niche C^2 will exceed the current state of the art. Advances will be most significant in the areas of processing, fusion, and encryption. Due to its reliance on civil systems, niche C^2 will be delayed relative to our own. It might also present single-point failure nodes and a hierarchical planning and tasking process.

It's a near-certainty future niche competitors will field *stealthy cruise missiles*. They are currently under development by a wide variety of sources. Any nation with a moderate defense budget should be able to buy several thousand stealthy cruise missiles capable of strike, communications relay, and surveillance. Therefore, we must assume at least a portion of the enemy's aerospace weapons will present low signatures.

A niche enemy will have a large inventory of *precision weapons*. Reflecting at least mid-1990s state of the art, these weapons will have less than 10-meter accuracy. They may depend on US-controlled navigation systems (e.g., global

---

*By 2002, experts estimate civilian imagery satellites will have one-meter resolution.

positioning systems). Some of these weapons will retain their accuracy regardless of weather or darkness.

Previous Systems

A niche's previous systems might consist of a handful of nuclear weapons, large stocks of chemical weapons, a limited number of ballistic missiles, and substantial numbers of late-generation traditional systems (e.g., tanks, aircraft, artillery, surface warships, mines).

A niche competitor would likely have a robust inventory of currently (mid-1990s) available weapons. These could include infantry, armor, artillery, submarines, mines, nonstealthy fighter aircraft, surface-to-air missiles, ASATs, chemical munitions, and short-range ballistic missiles (e.g., Scuds). The niche would use these previous systems to conquer territory, while using its emerging systems to combat US intervention. A niche would also have weapons of mass destruction, including nuclear weapons.

A niche competitor would probably have a limited number of nuclear weapons (less than 10). Without an intercontinental ballistic missile (ICBM), however, these weapons would not directly threaten US territory. Nevertheless, they'll threaten US allies and bases in the region. A niche could use these weapons to threaten a country which gives the US basing or overflight rights during a contingency. For example, a nuclear-armed North Korea should be expected to threaten Tokyo with a nuclear strike should Japan allow US forces to operate from Japanese bases during a war in Korea. Similarly, the North Koreans might inhibit a large US logistics flow by threatening Pusan with nuclear attack. In these examples, a niche would need only a handful of nuclear weapons to complicate US operations.

Putting all of these factors together, a niche competitor of 10–20 years from now will present challenges of a different nature from those posed by an MRC-scale competitor today. A future niche will be able to detect large US force deployments and relay this information to stealthy weapon systems. These systems will likely penetrate US aerospace defenses in significant numbers. Once in the target area, they'll strike with great accuracy. This combination of previous weapons (tanks, a few nuclear weapons, submarines, ASATs, surface-to-air missiles, etc.) plus emerging weapons (stealthy cruise missiles, civil satellites for reconnaissance and communications, etc.), orchestrated by a new CONOPS, would confront US aerospace forces with a demanding situation.

Asymmetric Employment Schemes

Further complicating this environment would be the likelihood of asymmetric employment schemes.* A niche competitor would likely avoid a direct confrontation with the US. Rather, the niche would attempt to offset US strengths by employing an indirect strategy. For example, a niche competitor would not seek information dominance. It is highly unlikely a niche will surpass the US military in information technologies over the next 10–20 years. US companies and universities lead the world in information technologies; the US military totally dominates the military applications of these technologies. Therefore, the niche might pursue an "information neutral" environment. It would attempt to "level the playing field" by degrading US information flows.

Information leveling could be accomplished several ways. One way would be through hackers. The niche could hire any number of computer hackers to attack US information networks. These hackers could be hired at the last minute, assuring state-of-the-art competence. They could be hired in large numbers from around the world; India and Russia, for example, have a wealth of software talent willing to work for relatively low pay. It would be very difficult for the US to assess the scope and direction of this campaign in advance of hostile intrusion.

---

*Conflict is almost always fought in an asymmetric manner. Even among peer competitors, symmetric conflict is rare. For example, Great Britain and Napoleonic France were peer competitors, even though the former relied mainly on its navy while the latter emphasized its army. The World War II Battle of the Atlantic between US/UK and Germany matched merchantmen/aircraft/destroyers versus U-boats. In the sports world, two football teams may be evenly matched although one emphasizes a running game and defense while the other relies on a passing attack. In conflict, asymmetries are the norm. When we use the term "asymmetric," we're trying to identify fundamentally different ends and means.

Fortunately, such an offensive has significant weaknesses. For example, the US could take steps to protect its vital information systems. Just as banks and businesses protect their information systems through encryption and protocols, the US military would use similar methods to protect its information systems. Another weakness is that disorganized hackers would probably bring little orchestration to their attacks. Lacking proper training and positive control, they would likely "service" targets with little regard to operational art. Finally, hackers would get little feedback on success or failure. They would not know whether they were successful; nor could they be sure they were entering a real system. Despite these weaknesses, however, hackers in the employ of a niche could pose a credible threat to US information systems. Serious defenses are mandatory.

A second approach to information leveling would be by physical attacks on US collection and communications satellites. The niche could launch primitive ASATs against these platforms, particularly those in low earth orbit. The niche could also detonate a nuclear weapon in space or in the upper atmosphere. The resulting electromagnetic pulse (EMP) would disable unshielded satellites. Replacement satellites, if also unshielded, would quickly degrade due to the enhanced radiation retained by the Van Allen Belts.[2] While niche systems in space would also be affected, EMP blasts would probably adversely affect US information forces—and thus, US operations—to a greater extent than those of the niche aggressor. An orchestrated campaign with ASATs and EMP blasts could degrade US space systems and cripple US military operations worldwide.

A third asymmetric employment strategy available to a niche is projection denial. Niche competitors will *not* have the means to conduct a long-range campaign against US forces with a high confidence of success. They would lack the NRT intelligence and manned penetrators necessary for such a campaign. However, niches could offset these shortcomings by combining a mix of relatively low-tech systems and weapons to make US power projection operations difficult, or even unfeasible. For example, a niche could mix a handful of modern diesel submarines and mine barriers to slow and canalize US sea lift. It could observe the resulting choke points with commercial overhead imagery then target specific ships with stealthy cruise missiles.

In this example, the niche would avoid challenging American naval forces directly. Its objective would not be command of the seas, but rather sea denial. The niche would use information obtained through third-party COMSATs to plot the movement of US forces at sea. It would then launch missiles towards anticipated LOCs. These missiles would use seekers with broad-area terminal guidance.

Should the US decide to absorb these attacks and remain in the war, it would face a decision. The US could either operate under this type of observation *or* attempt to interdict the niche's information. The latter would prove difficult in the Information Age where multiple sensors outside government control are available.* Data—and its means of transmission—is becoming ubiquitous. It seems most likely the US will be forced to operate under a limited amount of enemy observation. Prudent aerospace planners should allow for this probability.

By employing innovative operational concepts and a limited number of emerging weapons, a niche competitor could pose significant challenges to US operations in certain circumstances. Asymmetric employment concepts, particularly in the areas of information and power projection denial, might "level the playing field" to the point the US is dissuaded from involvement. To deal with this challenge, US aerospace forces should prepare to employ the following 10 operational concepts:

- Paralyze enemy command and control.
- Dominate battlefield awareness.
- Integrate space-based systems and unmanned aerial vehicles for conflict surveillance.
- Support the information campaign.
- Attack enemy wealth.
- Attack enemy invasion/occupation forces.

---

*Third-country satellites could be vulnerable to (1) government-to-government pressure, requesting the third-party state cease providing satellite information to the niche competitor, (2) to electronic warfare against the satellite or ground station, and (3) direct ASAT strikes.

- Establish aerospace superiority.
- Avoid deployment of critical fixed targets within range of enemy stealth.
- Airlift forces and logistics into the combat area.
- Support the ground counteroffensive.

Paralyze Enemy C^2

Every military professional knows the absolute necessity of continuous control of military operations. Murphy's Law is alive and well in the military profession; military operations as simple as change-of-command ceremonies require constant massaging. Wars are infinitely more complex than ceremonies. Unexpected obstacles and opportunities are the norm; no operation ever goes according to plan. As a result, each and every military operation requires an incredible amount of hands-on control. Continuous adjustments are *always* required. These adjustments depend upon positive command and control. Decision makers and the means to transmit their decisions make up this system. Thus, nodes crucial to this system are high-value targets. Their disruption would cascade chaos among subordinate units.

For a niche competitor, command and control nodes are a major vulnerability. Modern US surveillance systems (especially electromagnetic intelligence) are expert at identifying command links. Open-source literature can also provide wiring diagrams of communication flows. Once identified, these nodes are vulnerable to US attack.

There is little a niche competitor can do to forestall this vulnerability. If the niche constructs a command and control network comprised solely of hardened, indigenous systems, it will be, at best, rudimentary. The US could easily operate within the niche enemy's decision loop. If, at the other extreme, the niche uses world-class communications systems and protocols, it will expose itself to massive information interdiction. This interdiction could be remarkably precise. The US would pressure companies based within the US or allied nations to provide source code and architecture information. This knowledge would facilitate an information interdiction campaign. Unfortunately for the niche competitor, there is little recourse. Any information system, indigenous or imported, will have substantial drawbacks.

If the niche has indigenous satellites capable of NRT operations, they should top the US target list. NRT information architectures put all aspects of US military operations at risk. On the other hand, architectures using another country's civilian imagery satellites will be a less immediate threat, necessitating a less immediate response. Third country civilian systems are inherently less responsive than indigenous military systems. It will take time to get the right picture from the civilian satellite through the downlink station to the niche's operational headquarters for analysis and tasking. It may take days to produce an executable product. During the processing time, the US could redeploy forces or defenses.

The US will strike enemy C^2 nodes at each level of war. At the strategic level of war, national political and military leadership will be attacked. The goal will be to isolate the enemy's national decision makers from their instruments of power. These instruments may range from weapons of mass destruction, to air defense, to intelligence, to political control over their population. The niche's nuclear weapons should not sway this strategy. As long as the US effect is to isolate the enemy leadership from its means of command—as opposed to decapitation—the US can avoid placing enemy leadership in a suicidal corner.

At the operational level of war, field commanders will be severed from their subordinate units. At the tactical level, units will be cut off from their battle managers. Threat warnings and targeting information will arrive too late to do any good.

Dominate Battlefield Awareness

It is well within the realm of technical possibility to observe practically everything of operational significance about a battlefield. Admiral William Owens, Vice Chairman of the Joint Chiefs of Staff, called this concept *dominant battlefield awareness*. This concept has three components. First, platforms continuously surveil the area of interest. A mixture of aircraft, satellites, and UAVs, equipped with multispectral sensors, establishes 24-hour, all-weather coverage of the battle area. Unattended ground sensors sniff/watch/listen/ report along areas of possible maneuver. SOSUS-type sensors listen for underwater threats. Second, data generated by these sensors are fused and filtered through wide-area automatic target recognition software. This software cues more refined systems to specifically identify emitters and high-signature targets (e.g., armored formations or logistics points). Lastly, this information is disseminated to weapon systems. This dissemination takes advantage of large bandwidth and digital compression technologies. It transmits via direct broadcast satellites. The result of these three steps is dominant battle-field awareness.

A note of caution: Dominant battlefield awareness does not mean *perfect* knowledge of all enemy locations and intentions. Knowledge of *everything*, distributed to *everybody*, is impossible to attain. Plans based upon such an impossible standard are doomed to failure. Rather, dominant battlefield awareness is *an attempt to exploit order-of-magnitude increases in what's identifiable about a battlefield.*

Throughout military history, what commanders have not known about an adversary has dominated our image of war. The armies of generals Lee and Meade *bumped* into each other at Gettysburg. In 1914, the German army didn't even know the British Expeditionary Force was on the continent until they ran into the BEF at Mons.[3] Hitler kept panzer divisions in reserve near Calais, waiting for the "real" cross-channel invasion. Saddam's army had little knowledge of Gen H. Norman Schwarzkopf's deployment to the west for the "left hook." In each of these cases, commanders had little information on whole armies maneuvering in front of them. In today's Information Age, such military ignorance is *impossible* if one fields an integrated mix of sensors, filters, and disseminators—and protects this architecture from effective enemy interference.

Integrate Space-based Systems and UAVs for Conflict Surveillance

Niche competitors will probably have the ability to target satellites in LEO and to effect an EMP burst in space (via a nuclear explosion). The US must have ready counters for these probabilities. Of the two, the EMP threat may be the lesser challenge. Satellites must already be hardened due to solar activity. This shielding could be intensified to negate EMP effects. However, this shielding would be required on all satellites for which military operations are dependent—including civilian-owned communications satellites.*

The ASAT threat depends on target orbit. Satellites in geosynchronous earth orbit (GEO) should remain relatively secure. During our planning period (2005-2015), it is improbable that any niche will have air-launched access to GEO. Putting an ASAT into GEO will require a powerful booster. For any niche, such boosters will require a fixed launch complex. Such a complex should not survive conflict initiation; the US would attack fixed space launch facilities on day one of the war.

On the other hand, LEO satellites would be reachable by air-launched (e.g., the F-15/ASAT) or mobile ground-launched systems. US options for decreasing this vulnerability will be few. Targeting air-launched ASATs prior to launch is not likely; they'll be difficult to identify among the hundreds or thousands of similar type weapons. The most effective measure will probably involve two steps: (1) maneuver the satellites and (2) destroy the niche's space-tracking capability. Without solid data on satellite tracks, the niche would find targeting extremely difficult. Other defensive measures could include in-flight interception of the ASAT, stealthy satellites, and a rapid replacement capability.

What this means is that satellites would probably be targeted to varying degrees. Because communications satellites are primarily in GEO, they should survive (unless the niche develops a mobile ASAT with GEO range). Reconnaissance satellites, however, operate primarily in LEO. Their survival in war with a niche is less certain.

---

*The US could promote such shielding by reimbursing COMSAT owners for carrying the extra weight needed for shielding. The mechanism could parallel that presently used for the civil reserve air fleet (CRAF). The US government pays airlines for the extra weight some aircraft carry in order to be suitable for military operations.

As a result, the US must augment space-based systems with atmospheric systems. Fortunately, UAVs, along the lines of Tier II+ and Tier III-, are well along in development. High altitude-long endurance UAVs, with loiter times of 48–72 hours, are probable in our planning time frame. They promise sufficient loiter times and survivability to accomplish the surveillance mission. While UAVs have capabilities that recommend them in their own right, they are also necessary to provide redundancy for space-based systems. Satellites in LEO with predictable trajectories are simply too vulnerable to interdiction.

This technological solution brings with it an organizational challenge. The theater commander must integrate two fundamentally different architectures. The commander in chief (CINC) must integrate both space-based *and* atmospheric sensors and relays; total reliance on one or the other for critical tasks is not wise. One or the other may be unavailable to perform a specific job at a crucial time. It will be far more effective to integrate their tasks so that one type of system is not the sole source for any vital node. It will also be far more efficient to integrate the data acquired by each system. Because space-based and atmospheric systems are presently controlled by separate commanders (USCINCSPACE and the theater CINCs, respectively), this integration will require adjustments to current command relations.

Support the Information Campaign

Dominant battlefield awareness also requires denying the enemy a similar amount of information. As with a peer enemy, the theater CINC will task the Joint Force Information Component Commander to orchestrate a denial/distortion campaign. In a sense, the JFICC will deprive the information age to the niche. Once that's accomplished, the other components will end the industrial age.

Against a niche competitor, we should expect the JFICC to conduct a short, high-tempo information campaign. This is due to two factors. First, the niche's information target set would be smaller than that of a peer. By definition, a niche would be less robust in information infrastructure. Second, because the niche would not pose a likely nuclear threat to the US, fewer political restrictions on homeland attacks will come into play. This would permit attacks with all types of conventional munitions across all target categories. The result should be a short, intensive campaign on a limited number of targets in the most efficient way possible.

Aerospace forces would directly support this campaign. In most cases, targets should be highly precise. They would include connectivity (e.g., fiber-optic lines and radio/cellular antennas), nodes (e.g., switching stations), repair assets, downlink stations (e.g., satellite ground stations), fusion centers, and C^2 personnel. Munitions used against these targets will cover the gamut of the inventory—earth penetrators, MHD, EMP, CBU, HE, etc. Bombers and cruise missiles would serve as delivery platforms.

Attack Enemy Wealth

To undercut the niche's ability to continue the war and to punish it for starting the war in the first place, the CINC would probably direct aerospace forces to attack the niche's wealth. If the niche depends on trade, aerospace forces would identify and interdict that trade. Shipborne trade would be easiest to interdict; overland the most difficult. The goal would be to stop all *substantial* trade. Minor amounts of imports and exports would always occur (if nothing else, these would take the form of smuggling). But nation-supporting trade, dependent on large and regular flows of goods, is easily identified and stopped.

If the niche depends on imported resources (such as oil), trade in that resource could be struck as outlined above; targets could include trucks, bridges, shipping, ports, and trains. Also, aerospace forces could attack the supporting industry and its infrastructure. Those targets could include pumps, pipelines, refineries, storage tanks, and ports. Electrical production and transmission facilities crucial to this industry could also be attacked, as could banking and communications links with the rest of the world. Any of these approaches could cripple trade.

A service-based economy would provide an extremely vulnerable target set. Because services are readily substituted, an attacker need only *disrupt* transactions to cause customers to look elsewhere. For example, large customers can readily choose banks and insurance companies anywhere in the world. They pick specific firms based on convenience, reliability, and returns. Attacks on either the buildings housing such businesses—or their supporting electricity and communications—would quickly interrupt transactions to the point that customers would take their trade elsewhere. Because service economies are severely affected by disruptions, aerospace attacks against their infrastructure offer high leverage.

In those cases where crucial supporters of the niche government have commercial interests, attacks would focus on those interests. It does little good to impoverish common citizens when they have no voice in government policy. It makes great sense, however, to target the wealth of key government supporters. Intelligence agencies should provide precise insights into the commercial concerns of key regime supporters. Aerospace forces would then attack these enterprises.

Attack Enemy Invasion/Occupation Forces

Wars of the twentieth century have proven infantry's inability to overcome modern defenses by itself. Massed infantry charges didn't work at the Marne and haven't worked since. The Iran-Iraq War proved this lesson again. To overcome modern defenses, armies need armored mobility. They need tanks, armored personnel carriers, mobile artillery, and trucks—by the thousands. This means, by definition, that attacking armies bent on invasion and occupation are high-signature enterprises.

The combat forces themselves would also present a lucrative target set. US Defense Department planning guidance in 1993 described notional niche invaders as having at least 5,000 armored vehicles and several hundred thousand troops. [4] Such massed formations of tanks, troop carriers, and mobile artillery—necessary for all but a dispersed, footborne invasion—are readily detected. Once detected, they are vulnerable to aerospace attack. Bombers and cruise missiles, carrying a wide assortment of precision munitions, have a proven ability to destroy massed, slow-moving surface forces. Practically the

Armored Battalion in Tactical Road March

UK 1st Armored Division Entering Iraq During *Desert Storm* Ground Offensive

entire family of aerospace munitions under current development (sensor fused weapons, wide-area munitions, brilliant
antitank munitions) is optimized for this target set. Equipped
with advanced munitions either in service or about to become
operational and directed by modern C^3I systems, airpower has
the potential to destroy enemy ground forces either on the
move or in defensive positions at a high rate while concurrently destroying vital elements of the enemy's war-fighting
infrastructure.[5]

Niche competitors would be unable to rapidly replace lost
armored forces. Their inventories are limited and, because
niche regimes use armored forces for internal security and
border defense, they can ill afford to lose several thousand in a
war with the US. Also, once these inventories are cut,
replacements are difficult to obtain. Although used main battle
tanks and armored personnel carriers are available on the
world market, their quality is suspect. In addition, used
armored vehicles aren't cheap, spare parts availability is
uncertain, and hard currency is required for purchase.

New procurement is a poor alternative. Prices of new
armored vehicles are high and worldwide production is low. A
niche can't rely on the international arms market to produce
large numbers of armored vehicles. The capacity simply
doesn't exist in the post-cold-war world.

For these reasons, once the niche loses its armored forces, it
will have great difficulty replacing them. This difficulty will
cause internal and regional security problems for the niche.

In addition to their combat elements, invading/occupying
armies require immense amounts of support. Logistical "tails"
are vital to land force invasion/occupation. For example, a US
armored division consumes 600,000 gallons of fuel per day
when on the offensive.[6] In addition to the personnel and
transportation units assigned directly to each division, an
attacking force of six US divisions requires a support force of
roughly 200,000 personnel and 40,000 trucks.[7] These support
forces are mandatory elements of military operations. Simply
stated, armies don't move without logistics.*

---

*Even if an army could cut its logistics requirements by 50 percent, a substantial
number of critical targets would remain.

These logistics lines and stores are ripe for interdiction. They have high signatures. By definition they are less defended than combat forces. They move slowly and have few defense systems. As Eliot Cohen writes, "The weakest part of any military organization is its logistical train—the array of ill-protected trucks, storage depots, maintenance areas, and supply points that are needed to sustain forces in combat."[8] Precision munitions delivered with an element of surprise against enemy logistics should have a devastating effect. A major goal of US aerospace forces will be to "hollow out" an attacking army by gutting its logistics.

While a niche competitor's armored forces will not mirror US armored divisions, we can assume they will depend upon large amounts of support personnel and infrastructure. Even dividing the US numbers by a factor of two or three still leaves a substantial support force. That is the nature of maneuver units; large supporting elements are crucial.

US aerospace forces would concentrate on countering military invasions by striking an invader's maneuver forces, logistics, and C^4I from the outset of hostilities. This would not be an attrition campaign, however. The overall idea would be to deny an invader mass and maneuver, thereby forcing the enemy into a frontal attack scheme. US aerospace attacks on enemy armor, logistics, massed formations, C^4I, and air support would limit the enemy's offensive options to poorly coordinated infantry and artillery attacks. As noted above, this style of frontal assault is readily defeated by modern defenses. Should the enemy insist on massing its surface forces despite the threat of precision attack from aerospace weapons, wide-area munitions delivered by aerospace forces would have a decisive effect.

Establish Aerospace Superiority

We should expect niche competitors to field a limited number of ballistic missiles; many already do (e.g., Scuds). The speed, range, and survivability of mobile ballistic missiles make them attractive. Planners must incorporate the ballistic missile threat within their calculus when devising future aerospace superiority regimes. Fortunately, ballistic missiles in the hands of a niche competitor should have several major weaknesses.*

One weakness is that ballistic missiles offer a high signature. They emit a large infrared signature at launch and are radar reflective in all phases of flight. These high signatures mean that space-based systems can detect ballistic missile launches with high confidence and can then cue radar trackers. Once tracked, the minimal maneuverability of ballistic missiles makes them vulnerable to interception.**

Another weakness is that ballistic missiles are expensive. A niche would probably have fewer than a hundred. For example, Iraq launched 88 Scuds during the Gulf War.[9] Unclassified sources estimate North Korea's inventory of Scud launchers at 30; the Saudis have a total of 40 ballistic missiles.[10] It is unlikely a niche could afford the massive salvoes of ballistic missiles needed to overwhelm robust theater defenses.

A third weakness is that mobile targets are almost invulnerable to ballistic missiles. Unless the missile is equipped with brilliant sensors, the combination of missile inaccuracy and target movement makes for a very low Pk.

Lastly, while ballistic missiles are optimized for attacking targets with limited windows of vulnerability, this capability demands an extensive support network. This support network must include surveillance sensors, sensor-to-warhead target data transmission, and an NRT decision cycle. Such a network adds to unit costs and presents a lucrative target set for US attack.

Despite these vulnerabilities, the threat posed by ballistic missiles demands a layered and robust defense. US defenses should target enemy ballistic missiles throughout their deployment and employment. This means (1) disrupting ballistic missile C[4]I, (2) targeting ballistic missile launchers

---

*One weakness of current ballistic missiles that should not persist is inaccuracy. Currently, the CEP of ballistic missiles available to niche competitors is measured in hundreds of meters. These ballistic missiles are most suited for attacking wide-area, soft targets (such as cities). Their inaccuracy makes hardened point targets invulnerable to ballistic missile attack. Given expected improvements in precision and navigation technologies, this shortfall should end long before 2010.

**Improved aerospace technologies (e.g., lasers, kinetic kill) integrated with improved computing technologies offer considerable promise for high-confidence interceptions.

prior to launch, (3) detecting launches, (4) intercepting the missile/warhead in flight,* and (5) degrading the en route accuracy of the missile or terminal accuracy of the warhead. Total success in any one of these tasks, or substantial success in them all, will effectively defeat a niche ballistic missile attack.

Manned aircraft in the hands of a niche competitor is probably the easiest aerospace defense task. During the time period of this projection, the US will face only nonstealthy fighters. It's unlikely that a niche would have stealthy aircraft. The F-22 air superiority fighter won't even arrive in US inventories until the turn of the century. No other stealth air-to-air fighter is under serious development. This lack of stealth availability means niche air forces must rely on nonstealthy fighters. These fighters can't compete with the F-22. Thus, any niche confronting the US will find itself with a significant quality disadvantage in terms of air-to-air fighters.

Photo courtesy of US Air Force

Iraqi MiG-25 Destroyed in its Hardened Shelter by a Precision Bomb During *Operation Desert Storm*

---

*Boost phase, post-boost phase, or terminal.

Any niche air force opposing the US would also find itself at a serious quality disadvantage (see table 4). Even relatively large niche air forces, such as the Israeli, Indian, and Saudi air forces, possess less than 200 frontline fighters.* That's less than 5 percent of the US inventory of frontline fighters. Even assuming a 1:1 exchange ratio, any niche fighting the US would quickly lose its frontline fleet and pilots without any hope of quick replacement.

Table 4

Trends in Air Power

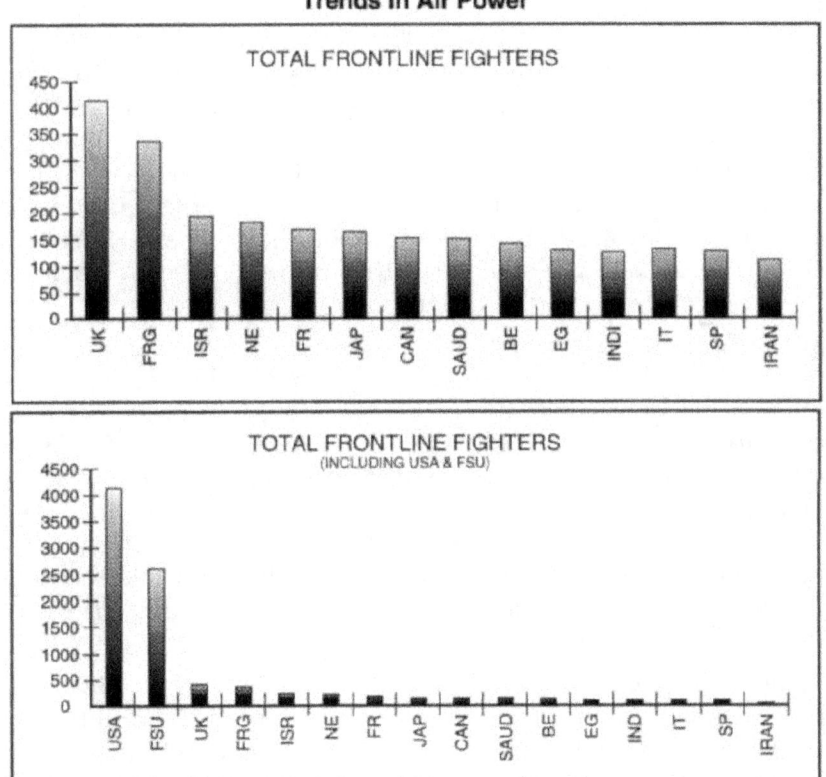

Source: *Trends in the Global Balance of Airpower*, RAND[11]

---

*Frontline fighters: F-14/15/16/18/117/, AV-8, MiG-29/31, Mirage 2000, Su-24/27, Tornado.

As we look to future war, the number of frontline fighters is a critical measure. Older fighters will be increasingly difficult to maintain. Old aircraft will become obsolete. How many modern aircraft a niche has today is the best indicator of future capabilities.

Finally, a niche's nonstealthy forces would be inadequately supported by EW, air-to-air missiles, tankers, training hours, intelligence, maintenance, etc. These shortfalls would make the niche's manned aircraft capabilities quite vulnerable.

The type of aerospace infrastructure necessary to compete with US aerospace forces is simply unaffordable for niche competitors. They would come to fight with an inferior position in each element of the air-to-air equation. An offensive counterair campaign against enemy air bases should be expected to destroy the niche's ability to launch coordinated strike packages. The few numbers of inevitable leakers should prove manageable by defensive counterair assets. Again, an effort will be required, but the components for an effective defense against a niche's *nonstealthy* manned aircraft during the next one or two decades are already programmed.

This optimism does not extend to unmanned atmospheric systems. The US will face a substantial challenge against stealthy enemy cruise missiles. Because cruise missiles require little in the way of infrastructure support, an offensive counterair campaign against cruise missiles would pay few dividends. For example, there would be few cruise missile "bases" to attack. The niche could launch cruise missiles from mobile TELs garrisoned in warehouses. It's not likely our intelligence agencies would find many cruise missiles prior to launch.

Further complicating cruise missile defense is the ability of cruise missiles to change direction. The inherent agility of cruise missiles makes "point defense" meaningless. *At no time prior to impact can the defender be certain of the cruise missile's target.* Although the cruise missile may seem headed for an unimportant area during one phase of flight, it might change direction towards a vital target during its next flight segment. This uncertainty stresses defenders. It puts defenders in the far more difficult "area defense" mode. Rather

94

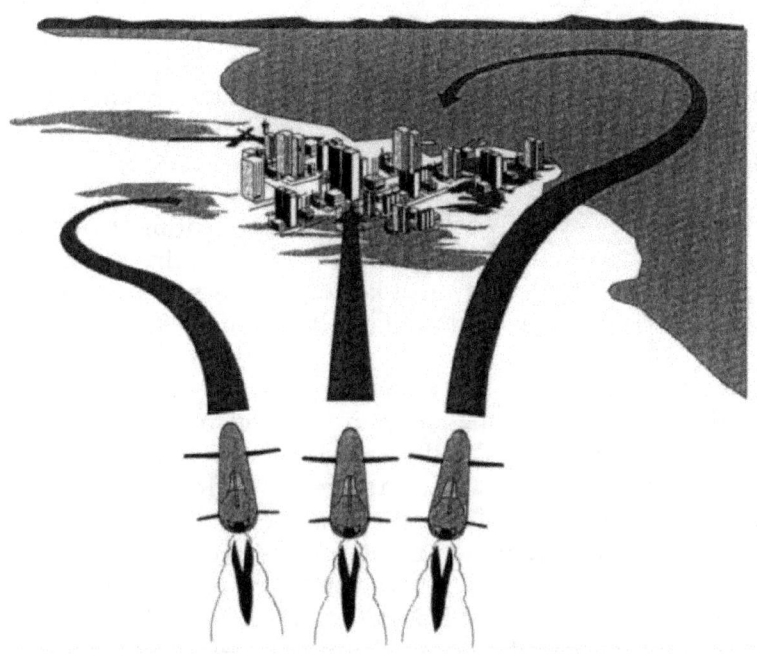

Figure 6. Cruise Missiles, Due to Their Ability to Change Course, Make Point Defense Difficult

than narrowing their efforts to point targets, they must defend large areas against precision attack.

An additional complication to area defense is the possibility that any one cruise missile might be carrying a nuclear warhead. Because even one nuclear impact could have strategic consequences, "zero leakage" becomes the standard. Not even one cruise missile can be allowed to penetrate, even when its suspected target is benign. This combination of zero leakage for an area defense makes cruise missile defense an extraordinarily demanding task.

In the cruise missile defense equation, C^4I may prove the most lucrative target, especially at the strategic level. C^4I for cruise missiles is a complicated task. The enemy must garner information on US vulnerabilities, decide on a course of action,

95

distribute taskings, then assess results. Disrupt any one leg in this structure and you degrade it all. Because C^{4}I supports each leg of this structure, it forms the key target set. Continuous attacks on the niche's C^{4}I should drive it towards uncoordinated and poorly considered use of the niche's cruise missiles.

The majority of the countercruise effort would be in defensive counterair. A mix of long-wavelength radars to detect stealthy cruise missiles,* high-powered targeting radars to refine the cued information, and large numbers of interception missiles** (both ground- and air-launched) offers some promise of partial air defense. The JFACC would need to tightly integrate these components to achieve an effective system. If the stealthy cruise missiles depend on off-board navigation aids like GPS, the US should interfere with the weak GPS signals. These defensive measures would significantly decrease the effects of niche cruise missile attacks.

Having said that, numerous stealthy cruise missiles will almost certainly penetrate even the most robust defenses. Therefore, *it's doubtful the US will be able to establish air supremacy if the niche has a substantial inventory of stealthy cruise missiles employed from mobile launchers.*

Avoid Deployment of Critical Fixed Targets within Range of Enemy Stealth

As with the case of the peer competitor, the niche must be expected to possess stealthy cruise missiles. Unlike a peer competitor, however, a niche's stealthy cruise missiles would have several significant limitations.

First, a niche's cruise missiles would be limited in number. Lacking a state-of-the-art defense industrial base, a niche would have to buy its missiles (or at least critical components) abroad. Normal constraints on sending currency abroad would likely limit procurement. Second, dependence on foreign

---

*Long-wavelength radars would be placed aboard aircraft, as opposed to being ground-based or carried on aerostats. Ground basing restricts detection to the horizon. Aerostat basing is difficult when conflict has already begun (set-up time is extensive for aerostats). Aerostats also present the enemy with an immobile, high-signature, high-value target.
**When fully developed, directed energy weapons may prove especially effective.

suppliers may also limit inventory replacements after the war begins. The US may be able to either interdict resupplies or dissuade suppliers. Third, a niche's cruise missiles would lack sophistication. Without a production base, the niche wouldn't have the full expertise available to a producer. This lack of indigenous expertise would diminish a niche's ability to rapidly reprogram or modify its missiles after a conflict began. Similarly, a lack of expertise would inhibit the niche's ability to maintain the state-of-the-art status of its missiles if there were a significant time period between purchase and use. Fourth, a niche's cruise missiles may not have real-time connectivity. A first-rate cruise missile offensive capability requires more than just the missiles themselves. It also requires advanced information and C^2 systems, all of which are expensive and require extensive operational expertise. A niche probably wouldn't have the capability to continuously monitor the entire battle area, identify key vulnerabilities, transmit this information to decision makers, then disseminate tasking orders to missile batteries—all in near real time. These four limitations would limit a niche's ability to effectively employ stealthy cruise missiles against US forces.

Despite these limitations, high-signature, immobile forces would be extraordinarily vulnerable to a niche competitor's cruise missiles. A niche would eventually detect large, fixed facilities via modern surveillance systems, and would attempt to overwhelm defenses with large numbers of low-signature cruise missiles. Those missiles that penetrate would probably hit vital targets with precision. The sum of these capabilities is militarily significant. Air supremacy over ammunition ships needing days to unload, airlift aircraft needing hours to off-load and refuel, large air bases with "tent cities," air refueling aircraft parked nose-to-tail in the open, and so forth, might not be possible .[12] If we make the mistake of giving a niche enough time to orchestrate cruise missile attacks on high-signature, immobile US forces, we'll likely see the niche exact a considerable toll.

This prospect forces a basic decision on future US employment CONOPS. The question is, will the US attempt to operate inside or outside the range of emerging enemy weapons (such as stealthy cruise missiles)? In other words, will the US attempt to (1) build sufficient defenses to negate the threat posed by a niche's inventory of emerging weapons or (2) stay outside the range of these weapons while conducting operations? The latter case drives force structure requirements towards long-range and standoff systems. The former heightens the value of radically new defense systems and low-signature, dispersed operations.

As with most things, the answer lies somewhere in the middle. To the greatest extent possible, the US must keep its forces outside the range of enemy stealth. Understanding that out-of-range basing won't be possible for all forces, the US must adjust its equipment and CONOPS to operate under the threat of a stealthy cruise missile attack by a niche enemy.

The US can take several steps to minimize its force structure within range of niche cruise missiles. It can take advantage of modern communications and keep key command and control functions (e.g., component commanders) within CONUS. Because most communications "bounce" off SATCOMs anyway, and significant aerospace forces (e.g., bombers,

Photo courtesy of Tsgt Paul gentlemen. Page, USAF

Aircraft Parking Ramp at Sheikh Isa, Bahrain, During *Operation Desert Shield*

tankers) base outside the theater, the rationale for keeping key aerospace C^2 nodes (e.g., JFACC) in-theater is probably based more on culture than operational efficiency. The US could also emphasize long-range, highly survivable systems (e.g., stealth bombers) at the expense of short-range ones (e.g., F-16) that require additional support aircraft. Finally, the US could minimize its in-theater logistical departments to the greatest extent possible.* These steps would lessen the number of critical targets placed within range of niche stealth.

Airlift Critical Supplies and Spare Parts into the Combat Area

A major difference between a peer competitor and a niche competitor is the niche's absence of near-real-time sensor-to-shooter systems. When fighting a niche, we can assume there will be a delay in the cycle time necessary for the niche to detect a US vulnerability, make an attack decision, disseminate the tasking, and put a weapon on target (time of flight). As long as airlift ground times are shorter than this time loop, airlift operations can proceed.

However, if the niche invests heavily in sensor-to-shooter technology, airlift operations could be readily targeted. For example, if the niche has direct downlinks from multiple third-country surveillance satellites, plus a rapid decision system, it could launch against our aerial ports during their most vulnerable windows. Several airlifters with their arriving forces on a small parking ramp would make a perfect target for stealthy cruise missiles with cluster munitions. In addition, if arriving units were to delay before dispersing, they would give the niche a similar opportunity to target massed forces.

A niche could also launch cruise missiles against routine airway traffic. Cruise missiles with active seekers could fly along standard arrival routes and airways. Highly reflective civilian airliners (i.e., the CRAF) on predictable routes, devoid of threat warning receivers or active countermeasures, could incur heavy losses.

---

*The policy of two-level maintenance supports this logic.

GPS jamming would further complicate airlift operations. To degrade enemy cruise missile targeting, the US would almost certainly jam or spoof the civilian GPS signal ("CA" code) in the vicinity of potential US target areas. While US military forces would retain access to the encrypted military GPS signal ("Y" code), US civilian airlifters without access to the GPS Y code would find their GPS navigation equipment unusable within the theater of operations. Combined with the absence of terminal navigation aids (e.g., TACAN/VOR/ILS), only airlifters with access to the GPS Y code would be capable of instrument navigation (to include instrument approaches).

To deal with this possible environment, airlifters should plan for rapid onloads and off-loads within theater, access by civilian airlifters to the GPS Y code, and staggered, nonperiodic routings.

Support the Ground Counteroffensive

Regardless of the effect of aerospace attacks on enemy ground forces, it's likely the niche leadership would leave its forward invasion elements in place. Even if its troops are halted, cut off, and under continuous attack from the air, a wholesale retreat is unlikely. Granted, such a strategy is illogical. Once its ground forces are stopped in their tracks with no chance of sufficient resupply, a logical enemy leadership would acknowledge the hopelessness of its situation and sue for peace. Unfortunately, such intelligence is rare. Just as Hitler left von Paulus' Sixth Army at Stalingrad and Saddam left his conscript divisions to be run over in the Kuwaiti theater of operations, we must anticipate that niche enemies would leave their invasion forces in place regardless of the effectiveness of air strikes against them. If these forces occupy territory of interest to the US or its allies, friendly ground forces would eventually have to launch a counteroffensive to drive them out.

Although the inevitability of a ground counteroffensive is of little doubt, its character would be the focus of much debate. Some might recommend a heavy, multidivision ground offensive (similar to the Desert Storm ground war). The upside of this approach is the likelihood of few friendly battle casualties. By using overwhelming firepower, the US should be able to shred what's left of the niche's ground invasion/

occupation force. However, this approach also has a downside. It requires a massive buildup prior to G day. This concentration would become a lucrative target for the niche's cruise missiles . This approach is also time intensive, requiring many months to move hundreds of thousands of troops (with their hundreds of thousands of tons of equipment and supplies) to the area of operations. During this time, the enemy might adapt to the air offensive and devise an alternative strategy.

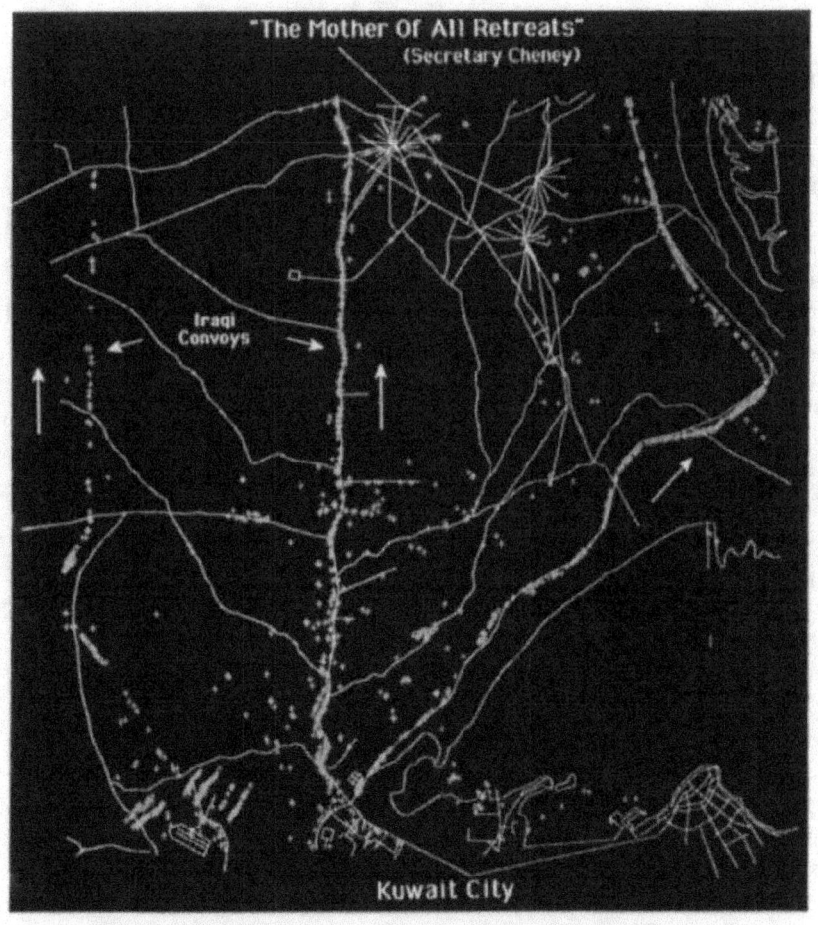

Source: Department of Defense

Figure 7. The Iraqi Army in Retreat from Kuwait City

A better approach would emphasize light ground forces for the counteroffensive. The source could be the local threatened ally, other US allies, and/or US Army and Marines. All would depend on airpower as the source of heavy fire support for maneuver units. This dependence would mean extensive taskings for aerospace forces.

Summary

This chapter discussed operational concepts for US aerospace forces in a future war with a niche competitor in 2010. In such a war, the United States would undoubtedly possess more and better aerospace weapons. The US aerospace arsenal would include stealth systems (cruise missiles and manned fighter-bombers), dedicated atmospheric and space-based information systems, automated C^2, precision munitions with submeter accuracy, nuclear weapons, and ballistic missiles with intercontinental range. A niche enemy would possess some of these systems in limited quantity. Because of the geopolitical environment, it's safe to assume the conflict would occur within and adjacent to the niche enemy's borders. These borders would be several thousand miles from the CONUS.

This war would be fundamentally different from that which is possible today. The biggest difference would lie in the inability of aerospace defenses to protect high-signature forces from attack. In a future war with a niche, we must assume the stealthy cruise missiles and aircraft of both sides *would penetrate aerospace defenses in significant number s*. After penetrating, these systems would target key vulnerabilities (due to modern surveillance systems) and would hit what they target (due to modern precision). Unlike Operation Desert Storm, large, fixed, high-signature US facilities would probably absorb attacks. This environment differs markedly from current conditions. It would dictate fundamental changes in our concepts of operation.

The following aspects of niche warfare deserve special emphasis:

• We must avoid deployments of critical fixed targets within range of enemy stealth. Their risk of destruction by stealth systems would be unacceptable.

• JFACC should base in CONUS. This fixed, permanent basing will allow immediate tasking of worldwide assets while excluding a high-value, high-signature target (JFACC HQ) from range of enemy stealth systems.

• LEO satellites would be lucrative targets absent extensive defensive measures. Satellites in GEO will be easier to defend; we can minimize a niche's limited GEO ASAT capability by attacking its space launch infrastructure.

• To undercut the niche's ability to continue the war and to punish it for starting the war in the first place, the CINC would probably direct aerospace forces to attack the niche's wealth. Targets could include trade, resources, and/or services. Assets of the ruling elites would have top priority.

• During the time period of this projection, the US would face only nonstealthy fighters. It's unlikely a niche would have stealthy aircraft. In addition, any niche would face a serious shortfall in numbers. Even assuming a 1:1 exchange ratio, any niche would lose its frontline fleet and pilots without any hope of quick replacement.

• The US must augment communications and reconnaissance satellites with unmanned atmospheric systems. UAVs promise sufficient loiter times and survivability to accomplish these missions. JFACC would forward NRT information on friendly and enemy maneuvers to allied forces. This transfer would require providing technical support and liaison officers.

• Degrading enemy cruise missile guidance would be a top priority. By manipulating external guidance systems such as GPS and by positioning decoys in the target area, defenders would attempt to exploit any algorithm weaknesses in the enemy system.

• Defenses would take advantage of ballistic missile vulnerabilities (large infrared signature at launch, radar reflective in flight, minimal maneuverability). Having said that, a 100 percent shield is probably impossible.

• We must mount a concentrated offensive against enemy C^4I. Cruise missiles and stealth bombers would assume this mission.

103

• Our defensive counterair campaign would emphasize sensor fusion. Enemy stealth systems would intermittently reflect and emit in flight, especially from the side and rear aspects. A thoroughly fused sensor network holds the possibility of successful detection and tracking.

• Aerospace forces would support the JFICC's campaign. All niche vulnerabilities would be targeted from the onset of the war. Aerospace taskings would likely be heavy.

• The aerospace campaign would attempt to deny enemy invasion/occupation, primarily through long-range bombers and cruise missiles delivering precision munitions.

• When airlifting critical supplies and spare parts into the combat area, we must minimize ground times. Depending upon distance from enemy missile launchers, ground times should be measured in terms of minutes, not hours.

• Aerospace forces would support counteroffensive forces. A counteroffensive would probably be necessary to reclaim territory from the aggressor.

• JFACC would attack the niche in parallel. All target types at all levels of war would come under attack near simultaneously. The goal of these attacks would not be attrition. Rather, the goal would be paralysis, especially of the enemy's C^2.

While not an exhaustive list, many of these aspects of a future CONOPS differ significantly from that currently envisioned for MRC planning. Accepting stealth as a given for both sides, we must figure out how to best operate in this environment. Signature reduction would be key to successful operations. Systems with small signatures have a high probability of survival. Those with large signatures would eventually be detected and hit. The implications of this environment deserve extensive examination and debate.

Notes

1. Paul Bracken, "The Military After Next," *The Washington Quarterly*, Autumn 1993, 158.

2. R. C. Webb et al., "The Commercial and Military Satellite Survivability Crisis," *Defense Electronics* 27, no. 8 (August 1995): 21–25. A low-yield nuclear detonation would likely "pump" the Van Allen Belts. The

enhanced belts would maintain a high radiation state for over a year, thus degrading unshielded replacement satellites.

3. Barbara Tuchman, *The Guns of August* (New York: Bantam, 1962), 280–81.

4. Les Aspin, *Report of the Bottom-Up Review*, October 1993, 13.

5. Christopher Bowie et al., *The New Calculus* (Santa Monica, Calif.: RAND, 1993), 83–84.

6. Field Manual (FM) 100-5, *Operations*, 1986, 60.

7. Bowie, 70.

8. "Two Types of Airpower," *Orbis*, Spring 1995, 190.

9. Office of the Secretary of the Air Force, *Gulf War Airpower Survey*, vol. 5, 1993, 562. (Secret) Information extracted is unclassified.

10. *The Military Balance 1994–1995* (London: International Institute of Strategic Studies, 1994), 137, 178.

11. Christopher J. Bowie et al., "Trends in the Global Balance of Airpower" (Santa Monica, Calif.: RAND, 1995). Data is current as of 1991.

12. It takes approximately nine days to off-load a break-bulk ammunition ship. Point paper N513, Headquarters USN, 25 November 1992.

Chapter 4

Near-Term Actions

This work is only a first attempt at building a vision of future aerospace warfare. A more accurate and comprehensive vision is needed. An institutional effort will be required for the US Department of Defense to build and implement this vision. Three initiatives would considerably promote this institutional effort:

1. Designate a focal point for future warfare. A vision of future war is important to today's decision makers; the fact one doesn't exist reveals a serious shortcoming. To permanently fix this shortfall, the Chairman, Joint Chiefs of Staff (CJCS), should elevate this responsibility to the director level on the Joint Staff. A Joint Staff Director would have sufficient clout to garner the necessary resources and attention to aggressively pursue a vision of future warfare over the long term.

One option would have J5 assume this responsibility—as its sole job. Released from any focus on day-to-day events, J5 would develop the future war-fighting vision for the US military and drive its implementation. Planning future warfare (mid- and long-term) would be J5's only job (regional affairs and arms control would move to J3).

2. Build a concepts development center (CDC). This center would have the sole, permanent mission of developing operational concepts for future warfare. Their initial focus should be 2010. Similar to the war-gaming efforts of the Naval War College in 1920–1930s, the CDC would conduct an ongoing series of war games. These games should generate new concepts of operation and supporting organizations for future warfare. It would develop concepts for all elements of the joint campaign, to include aerospace operations. Insights gained could build from one game to the next. CDC would also main- tain a clearing house for the insights learned from other Service war games centering on the future. J5 would drive this effort.

107

3. Build an information warfare center. This should be a new organization dedicated to developing war-fighting concepts for the Information Age. It should deal with all forms of i nfor-mation war (strategic/operational/tactical; offensive/ defensive; all missions). This center should have a mix of civilian, active duty, and reserve manning. J5 would also drive this effort.*

Although these groups will overlap in some areas, each should develop future CONOPS independently. Both should regularly brief joint and service leadership with stakes in future warfare.

Immediate Action

In addition to these broad efforts to clarify our vision of futur e war, five narrower actions are appropriate for immediate implementation.**

Task Defense Intelligence Agency for comprehensive future projections. The Defense Intelligence Agency (DIA) does a good job of estimating worldwide arms inventories and force structure trends. That effort should continue. However, future capabilities of possible adversaries involve more than weapons inventories. How a military intends to operate is also important. Therefore, DIA should also focus on trends in doctrine (as revealed by writings, HUMINT, exercises). This focus is necessary to discover the innovative operational concepts and organizational structures of possible enemies—which may differ radically from: (1) ours; (2) anyone else's; and (3) what we expect. Reviews of foreign initiatives can serve as a critique on US concept development. J5 will monitor this effort.

Study out-of-theater C². Our present CONOPS deploys C² to the theater of operations. The CINC and component commanders deploy close to the fight. This concept made sense when proximity to the battle was necessary to obtain accurate information. However, high-value C2 nodes within range of enemy stealth systems is an ever-increasing risk.

---

*Assuming that CJCS made J5 the focal point for future warfare.
**These action items need central management to ensure institutional follow-through. Each paragraph indicates J5 as that focal point (assuming CJCS designates J5 as the planner for future war).

Also, C^2 deployments delay development and implementation of campaign plans during transit and setup. Since today's joint C4I architecture makes accurate information available at great distances, it is now possible to command forces from greater distances. While great distances will make personal leadership more difficult, this is nothing new. It's already impossible for component commanders to see/visit all subordinate forces. For example, aerospace forces in the 1991 Gulf War were spread from Spain to Diego Garcia, from the Red Sea to the Persian Gulf. Some were even based in the CONUS. Finally, considerable manpower and facility support are needed to support each theater. For these reasons, we need to review component C^2 with an eye towards consolidation within CONUS. J5 (not J3 or J6) should review our C^2 architectures.

Study centralized control of national and theater collection platforms. National and theater surveillance systems are poorly integrated. They are tasked separately. Much of their output feeds separate databases. Because future enemies will target our information systems, these systems must function together. Should one go down, another must immediately take up the slack. This can best be done by a centralized command authority. CJCS should direct J5 (not J6) to report with specific recommendations within the year.

Organize for information war. IW has offensive and defensive aspects. It is conducted at the strategic, operational, and tactical levels. It has many subsets (e.g., jamming, viruses, psyops). Unfortunately, no one is responsible for bringing the art of war to IW in its entirety. NSA has a piece. So do SPACECOM, the services, each CINC, and so forth. Because information supremacy will be as important as air supremacy to future war, it's time to designate a CINC for IW. This CINC will work IW at the strategic and operational levels of war (components will retain tactical-level IW—just as they retain tactical aircraft while the AF works strategic and operational airpower). CJCS should designate CINCSPACE as the IW "king." J5 will monitor progress. If rapid progress is not forthcoming, CJCS could create another CINC or give the mission to one of the services as the executive agent.

Organize for parallel war. General Fogelman spoke of the possibility of attacking 1,500 targets the first day of a war. If his vision is correct, we're looking at far more than just an increase in efficiency—we're looking at the possibility of a new style of warfare. Future war may be conducted by attacking an enemy across all target sets and all levels of war near simultaneously. This type of war could strip an opponent of the ability to repair and adapt. Initial operations may prove decisive. CJCS should task the CINCs and services to brief the JCS on their appraisal of parallel war and its applicability to their mission (to include their vulnerabilities to this type of warfare). J5 will coordinate this review.

Required Flexibilities

Future warfare will also require specific flexibilities within weapon systems. Decisions made today will affect that flexibility. Therefore, today's acquisition considerations for aerospace forces should include these factors:

Information. Platforms must have the ability to incorporate/upgrade the latest information hardware and software, employ information obtained by off-board sensors, and transmit information garnered by onboard sensors to other weapon systems. Systems must also be able to operate despite a corrupted information environment.

Long range. Aerospace platforms should be based as far from enemy stealth systems as possible. Distance either puts a base out of enemy stealth range or gives layered defenses more opportunities to detect and target enemy attacks. Short-range systems will contribute only in very low-threat environments.

Stealth. High-signature aerospace weapons won't survive in future war. Weapon systems must emphasize passive sensing, minimal reflectivity, and discrete emissions. If platforms have these characteristics but their support structures (e.g., tankers, AWACS, fixed air bases) do not, the platform as a weapon system will not survive.

Precision. Manned aerospace platforms will become increasingly expensive. Driven by their need to incorporate long range, stealth, data processing, and mobility, there's no way they will also be cheap. This expense will drive down inventories. At the same time, target sets are expanding (better C^2 will allow dispersion; the possibility of strategic attack adds to the number of targets). There's also the desirability of conducting near-simultaneous attack across all levels of war. Precision is required to reconcile these trends. Each sortie must kill multiple targets.

UAVs. We need to think in terms of tens of thousands of UAVs. Their inherent stealthiness and minimal basing requirements allow low-signature operations. Their lack of an aircrew allows casualty-indifferent operations (although not collateral damage-indifferent). They are increasingly capable of long-endurance flights. They can perform strike, communications, and surveillance missions. While manned platforms will remain mandatory for certain types of missions (e.g., killing hard targets or targets with uncertain locations), UAVs will make decisive contributions to future aerospace operations if employed skillfully in large numbers. At the same time, future US defenses must have the capability to deal with large numbers of stealthy UAVs (i.e., cruise missiles) attacking simultaneously. Attacks by several hundred missiles at one time are plausible.

Mobility. One result of the Information Age will be the enemy's near-certain detection of fixed facilities. To offset this information, future commanders will need the flexibility to move land-based aerospace forces between bases. Such mobility requires a lean support structure. This concept affects how we envision munitions, C^2, maintenance, POL, and support equipment.

Alternatives to space. Satellites in fixed orbits will be exceedingly vulnerable in the future. Military operations dependent on satellite support rest on a dubious assumption of satellite survivability. We need alternatives to space-borne architectures. These alternatives should emphasize HALE UAVs and fiber-optic cable.

Power projection. Finally, the Information Age will fundamentally affect power projection. Ubiquitous sensors and transmission devices will give our future military commanders extensive information on the enemy's scheme of maneuver. Unfortunately, the enemy will also have substantial information about our forces. This information will make

111

either side's invasion forces exceptionally vulnerable when they mass to attack. Mobile defenses accompanying massed forces will be inadequate to stop interdiction forces emphasizing state-of-the-art information, C^2, penetration, and precision. It is at this point, very early in the battle, that wars will be won or lost. Once territory is seized, it may prove excessively costly to reclaim. Therefore, future US weapons must be capable of "day one" operations. US weapons must have the capacity to strike with overwhelming force from the first day of the war.

Appendix A

Illustrative Scenario (Peer)

To illustrate future war with a peer competitor, imagine the following hypothetical scenario. China invades the Maritime Provinces of Russia. China's announced goal is to foster an independent state; it promises to withdraw all troops after independence is established. China's Peoples' Liberation Army (PLA) immediately captures Khabarovsk, cutting the Trans-Siberian and BAM (Baikal-Amur Mainline) railroads. The remaining Russian units in Khabarovsk retreat to Komsomolsk. The PLA does not pursue, instead relying on air attacks and lack of supplies to neutralize the remaining Russian forces. The only other substantial Russian ground force in the area is holding Vladivostok, but is under heavy pressure. Russian and Chinese air forces battle overhead. Due to low sortie rates and effective SAMs, neither can establish air superiority. The UN Security Council meets, but China vetoes all measures. China issues a demarche on strategic attacks on Chinese soil: any attack, conventional or nuclear, on its soil will trigger a nuclear response on the soil of the attacking state. The Russians, unwilling to risk nuclear war, refrain from nuclear first-use. However, the Russians do attack Chinese military targets up to 200 NM deep into China.

The US decides to unilaterally enter the conflict on Russia's side.* The president orders the theater commander to achieve these five objectives:

- Reduce the impact of air attacks on Russian forces.
- Enhance Russian combat capabilities in the Maritime Provinces.
- Resupply Russian forces in Vladivostok and Komsomolsk.
- "Attrit" PLA forces on Russian territory (especially those attacking Vladivostok).
- Support the Russian counterattack from Komsomolsk towards Khabarovsk.

---

*This is only an illustrative scenario. The specific reason for intervening is not impor-tant for scenario purposes. What *is* important is to outline a contingency in which the enemy is a peer and the fight is thousands of miles from US territory.

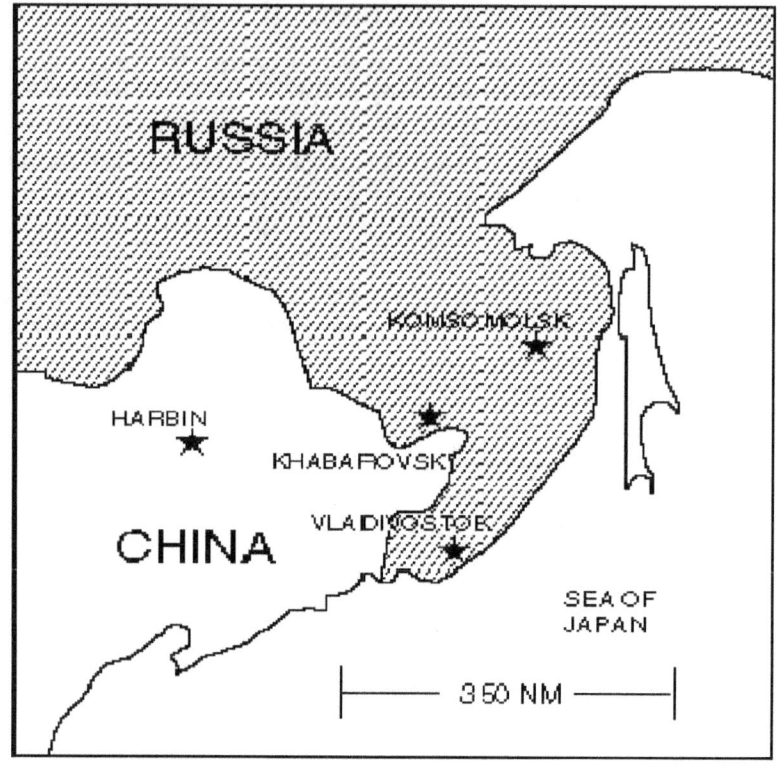

Figure 8. Map of Sino-Russian Area of Operations

Figure 8. Map of Sino-Russian Area of Operations

In accordance with this strategic guidance, the theater commander tasks his Joint Force Air Component Commander (JFACC) to accomplish five specific objectives:

• With the Russian air force, establish air superiority over the Maritime Provinces.

• Airlift US combat support and combat service support units to Komsomolsk and Vladivostok.

• Airlift Russian supplies into Komsomolsk and Vladivostok.

• Attack key Chinese military forces and LOCs on Russian territory.

116

• Provide near-real-time information on Chinese military movements to Russian commanders.

The JFACC faces several significant problems. First among them is China's considerable stealth inventory. The PLAAF has 100,000 stealthy cruise missiles and 1,000 mobile launchers.* Range: 1,000 NM. RCS: <.001 frontal and rear aspects; <.01 side aspect. Guidance: integrated GPS/INS en route; millimeter wave seeker in the target area. CEP is <1 meter. Warhead: 1,000-lb conventional, CBU, infrared cameras with data burst transmitters, multichannel jammers, decoys, or 100-KT nuclear. From launch areas around Harbin, Beijing, and Hong Kong, Chinese stealthy cruise missiles can strike all major East Asian cities (e.g., Tokyo, Seoul, Hanoi, Vladivostok).

The PLAAF also has 72 stealthy fighter-bombers divided among three bases. Their combat radius is 500 NM (unrefueled High-High-High Profile). Seoul, the Maritime Provinces, and the Sea of Japan lie within their range. Chinese stealthy fighter-bombers have an RCS between .1 and .001, depending on aspect. Each aircraft can carry either four long-range launch-and-leave air-to-air missiles or eight precision bombs (or canisters). Aircraft guidance: integrated GPS/INS en route. Warhead guidance is either visual (e.g., laser guided) or through pattern recognition sensors.

The JFACC's second problem is that he may *not* plan on using Japanese or Korean bases for offensive operations. When the Japanese government appeared ready to announce its active support for the combined Russo-US operation (including basing and cooperative air defense), Chinese bombers sank five oil supertankers approaching Japan. After the US Seventh Fleet seized Chinese merchant ships in international waters, Chinese stealthy cruise missiles struck two USN ships at Yokosuka Naval Station, Japan. Damage was heavy, casualties light. The Chinese also publicized their ability to hit nuclear power plants in Japan with cruise missiles.** As a result of these events, the

---

*China produces all missile components indigenously.
**Japan has 47 operating reactors. See "The Asian Syndrome: Many More Reactors," *New York Times*, 23 April 1995.

117

Japanese government restricted US Forces Japan (USFJ) to defensive operations only. The (unified) Korean government followed suit.

The JFACC's third problem is that China's space-based surveillance and communications systems can detect and relay information on large US buildups. China's on-orbit collection systems include NRT imaging, ELINT, and SBWAS

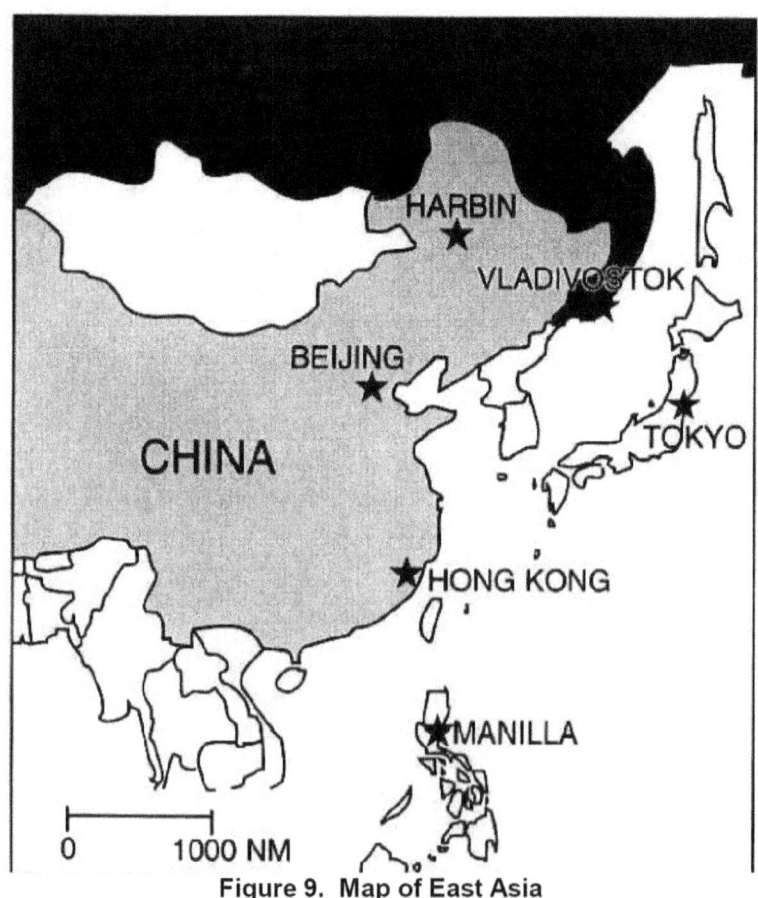

Figure 9. Map of East Asia

(radar). On-orbit communications systems include a Chinese-owned civilian constellation (serving Asia and Oceania) in LEO plus three military communications satellites in GEO. China also has replacement satellites and available launchers ready to go.

A fourth problem is that per NCA guidance, JFACC *cannot* conduct a strategic air campaign—even with conventional weapons—against critical targets on Chinese territory.* The NCA places leadership, C^{4}I, and weapons (including cruise missile and stealthy fighter-bomber bases) on Chinese territory off-limits to attack. However, the NCA *does* allow attacks with *nonlethal* weapons against these targets.

The JFACC's fifth problem is that the Chinese have SOSUS-type arrays in both the Sea of Japan and the East China Sea. The Chinese used commercial ships during peacetime to deploy this system. It is connected via fiber-optic cable with a control center in China (undetermined location). It is a relatively crude system; quiet submarines can operate with little chance of detection. However, large-screw surface ships and submarines firing missiles are readily detected. As a result, US Navy ships capable of offensive operations will assume a high level of risk if they operate in either the Sea of Japan or the East China Sea.

Table 5

Vladivostok Weather (Ceilings in Percent)

	Jan	Feb	Mar	Apr	May	Jun	Jul	Aug	Sep	Oct	Nov	Dec
Ceilings <3,000	9	11	23	33	38	71	68	52	33	28	19	11
Ceilings <1,000	4	5	12	19	24	57	53	33	13	10	7	4
Inches Precipitation	.3	.4	.7	1.2	2.1	2.9	3.3	4.7	4.3	1.9	1.2	.6

Source: USAFETAC/DOO (Now AFCCC)

---

*Similar to China's status as a sanctuary during the Korean War.

Weather works both for and against air operations. If the war starts in summer, low ceilings will hamper air strikes. Summer is the rainy season in the Maritime Provinces; ceilings are low, precipitation high. On the plus side, Chinese movements are restricted. Wet, marshy terrain would channel any invading Chinese forces along bridges and prepared roads. This channeling would make them easier to detect and attack. Conversely, if we were to assume a winter offensive, the opposite conditions would exist. Ceilings would rise and precipitation would decrease, facilitating air strikes on surface forces. However, frozen rivers and marshes would afford the Chinese more avenues of advance. For purposes of this scenario, the war starts in the summer.

In this scenario, China uses massive numbers of troops to invade Russian territory while stealthy aircraft and cruise missiles attack key Russian targets. These stealthy systems (cruise missiles and stealthy fighter-bombers) pose a significant threat to any deployment of US forces within 1,000 NM of China. Stealthy cruise missiles place soft targets (e.g., embarkation ports, air bases, aircraft carriers, nonhardened C^2 facilities) at significant risk during daylight, followed by either cruise missile or stealthy fighter-bomber attack at night. Chinese stealthy fighter-bombers could also attack high-signature airborne targets (e.g., AWACS, E-2C, J-STARS, C-17, B-1) operating within 500 NM of China. This threat slows down US deployments and limits the types of forces capable of effective participation.

US Aerospace Concepts of Operations

In this hypothetical scenario, China's stealthy cruise missiles have the ability to target US forces within 1,000 NM; Chinese stealthy fighter-bombers can target US forces within 500 NM. Both operate from a sanctuary enforced by the threat of nuclear weapons. US basing options are degraded by Japanese and Korean neutrality. US forces have access to bases in Russia, Alaska, the CONUS, Guam, and Taiwan.

In accordance with guidance from the theater commander, the US aerospace CONOPS contains the following major themes:

- Limit the number of US targets vulnerable to attack.
- Command US aerospace operations from Cheyenne Mountain.
- Protect US space systems.
- Attack China's satellites.
- Provide NRT information on the enemy to Russian defenders.
- Degrade Chinese stealth cruise missile accuracy.
- Attack China's stealth aircraft force.
- Defend against Chinese stealth attack.
- Defend against Chinese theater ballistic missile attack.
- Attack Chinese strategic and operational information.
- Deny Chinese invasion forces mass, maneuver, and effective C^4I.
- Airlift critical supplies to Russian forces.
- Prosecute a parallel war.

Limit the Number of Targets Vulnerable to Attack

Most US combat units remain more than 1,000 NM from China, outside the range of Chinese cruise missiles. The weight of the US combat effort falls on aerospace forces, primarily long-range strike aircraft and missiles. Short-legged, nonstealthy aircraft (e.g., F-16, F-18) have little use. Because of the presence of Russian ground forces in the Maritime Provinces, the US does not deploy ground combat units. The CINC limits deployments within range of Chinese stealth to "enhancing" capabilities: C^4I, air defense units (including spoofing and jamming teams), surface-to-surface missile batteries (e.g., cruise missiles, UAV launchers), and lift infrastructure. Those forces disperse after landing to preclude rapid identification and targeting (their approximate locations will have already been revealed by the high-signature airlifter). Units remain mobile and camouflaged.

Command US Aerospace Operations from Cheyenne Mountain

The theater CINC deploys forward; the JFACC and staff remain at Cheyenne Mountain.* The CINC gives general guidance (e.g., apportionment, overall objectives) through a liaison office to the JFACC. The JFACC deploys staff elements forward (to Vladivostok and Komsomolsk) to liaise with Russian forces (especially army units and ADA).

Fixed basing of the JFACC allows immediate planning/tasking of the air campaign. There's no delay in aerospace operations as the JFACC headquarters deploys to the Russian far east. There's also the crucial benefit of *not* positioning an important target within range of enemy stealth systems.

The JFACC at Cheyenne Mountain is colocated with the Joint Force Information Component Commander (JFICC). This dual headquarters receives all data collected by satellites, UAVs, aircraft, and surface sensors. This data is relayed through SATCOMs, backed up by COMSATs and three strings of HALE UAVs across the Sea of Japan. Each string consists of two UAVs separated by 500 NM. They relay data to separate ground stations in Japan which are connected by fiber-optic cable to the US grid. Data moves through the US grid to Cheyenne Mountain. As an additional safety measure, the JFICC orbits spare SATCOMs while readying more for immediate launch.

JFACC tasks strike aircraft regardless of location (e.g., stealth bombers based in CONUS, FB-22s at Petropavlovsk). JFACC directs employment of air defense units in theater, cruise missiles aboard ships at sea, surveillance aircraft in theater, and air refuelers in Alaska, Guam, and central

---

*Cheyenne Mountain has the advantage of proximity to USSPACECOM, putting two major aerospace component commanders together. It is also immune to all but a nuclear attack (barring development of kinetic rods from space). Other possible locations include Falcon AFB (also has excellent connectivity with USSPACECOM), Langley AFB (wealth of personnel with operational expertise), and the Pentagon (ready access to satellite downlinks, intelligence, political leadership). In all cases, significant parts of the JFACC's staff could be located elsewhere. For example, the JFACC could be at Falcon AFB with logistics coordinated from Langley AFB while strategic targeting is drafted by CHECKMATE in the Pentagon.

Russia.* It works closely with the colocated JFICC to manage and position surveillance platforms, and to deconflict communications requirements.

Unit taskings in the ATO are accompanied by the requisite intelligence information ("smart push"). Units can request specific additions ("warrior pull"), but the majority of their information is managed centrally. Smart push/warrior pull requires considerable expertise on the JFACC staff. Officers must anticipate the intelligence needs of operating units. However, it has the advantage of decreasing ATO cycle times while facilitating bandwidth management.

Protect US Space Systems

Having equipped its satellites with shielding against lasers and EMP, US satellite defenses are primarily concerned with Chinese jamming and direct ascent ASAT. To offset the jamming threat, the US relies on frequency-hopping EHF transmissions. Their focused footprint makes jamming ineffective. Nonlethal strikes on high-power jammers within China (readily pinpointed by their high signature), plus suppression of Chinese jammers within Russia, decrease interference on UHF channels. That leaves Chinese ASATs (and space mines) as the greatest threat.

At the outset of hostilities, China's two space mines in GEO drift toward US satellites. Due to their lack of either escorts or maneuverability, two US SATCOMs are destroyed. The US responds with ASAT strikes on all Chinese satellites, regardless of orbit. China follows suit. Both sides, however, refrain from attacking the other's ballistic missile warning satellites (e.g., DSP).

Despite the risk of nuclear retaliation, both sides attack missile launch facilities. With NCA approval, the CINC orders cruise missile attacks against the limited number of Chinese-fixed space launch facilities. The Chinese retaliate with cruise

---

*Since this scenario involves Russia as an ally, and Russia has world-class SAMs, there's no need in this particular scenario to deploy US SAMs (e.g., Patriot) to the AOR.

missile attacks against Vandenberg AFB and Kennedy Space Center (launched from "commercial" ships), plus similar Russian facilities. As a result of these attacks, all three nations lose their heavy space-launch capabilities. They can neither launch large satellites nor put sophisticated ASATs into GEO.

The US also attacks China's single space tracking facility. This attack significantly increases the survivability of US satellites. By maneuvering satellites and denying the Chinese the ability to identify the new tracks, the US impairs China's ability to precisely target US satellites.

Remaining US satellites in GEO survive. Space-based infrared systems detect air-launched ASATs upon launch, giving satellite controllers ample warning. Controllers use satellite maneuver and escort satellites to defeat the Chinese air-launched ASATs sent against GEO targets. In addition, the US attrits host aircraft for air-launched ASATs. Because these aircraft are nonstealthy and use tracks within five degrees of the equator for launch, they are readily detected and intercepted by AEGIS-type ships in the Indian Ocean.

US satellites in LEO also use maneuver to complicate Chinese air-launched ASAT targeting. However, maneuverability is limited due to fuel restrictions. As a result, the Chinese heavily attrit satellites in LEO. Constant maneuvering also degrades collection by these LEO satellites.

The loss of heavy launch complexes makes replacement of large platforms impossible. Replacement alternatives follow two paths. First, a fleet of 1,000 stealthy HALE UAVs carrying a variety of sensors assumes lost capabilities. This fleet launches from the area around Lake Baikal (Irkutsk, Ulan-Ude), the Kamchatka peninsula, and the Kurile Islands. Launch sites rely upon dispersion for defense. Second, air-launched boosters (similar to today's Pegasus) put attrition fillers into LEO.* Some of these satellites continuously maneuver until their fuel is exhausted, then remain in orbit until destroyed by Chinese ASATs. Others continuously maneuver until their fuel is exhausted, then

---

*Host aircraft operate from Florida. Launch tracks are along the equator (3,000 NM round-trip).

deorbit for recovery and relaunch. All satellites are single-mission and stealthy.

Attack China's Satellites

US attacks on Chinese satellites parallel China's space supremacy effort. US antisatellite jammers and lasers are based in space, in Russia, and in the US. Jammers in Russia, deployed by the hundreds, regularly move to complicate Chinese suppression attempts. Lasers attempt "soft kills" against Chinese satellites.

Russian special operations forces, along with US and Russian cruise missiles, attack Chinese satellite ground stations. US forces attack all Chinese satellite ground stations outside of China (e.g., ships in international waters).

Due to the loss of heavy space launch facilities, the US relies on air-launched ASATs. The US uses air bases in its southeast to launch host aircraft. These bases support both launches into GEO (from along the equator) and LEO insertions along polar tracks. The US also launches ASATs along HEOs to attempt intercepts with Chinese satellites in GEO.

ASATs are capable of carrying a variety of warheads. To reduce the chance of space debris, initial ASATs carry nonlethal warheads. They use EMP, MHD, and HPM for soft kills on Chinese satellites. For those satellites immune to soft kills, kinetic kill warheads on ASATs would follow. Given a robust inventory of ASATs with various warheads, the US could heavily attrit Chinese satellites, regardless of orbit.

Provide Near-Real-Time Information on Enemy Military Maneuvers to the Russian Defenders

High-value airborne platforms (AWACS, J-STARS, TR-1, Rivet Joint) launch and recover from bases >1,000 NM from China. Bases include Shemya, Petropavlovsk, Elmendorf, and Irkutsk.* Orbits within range of Chinese stealth fighters (500-NM radius) and cruise missiles (1,000-NM radius;

---

*As long as China maintains space-based surveillance systems and ICBMs, high-value assets would avoid using the same air base on consecutive nights. This mobility drastically increases support and C^2 requirements, plus crew ratios.

carrying antiradiation sensors) are possible, but risky. Stealth fighters escort surveillance platforms. However, given the importance of these platforms, their limited numbers, and their large signatures, attrition is high; they are priority targets for Chinese counterair weapons. For this reason, JFACC restricts their orbits to overhead rear area bases and convoys.

The JFACC integrates space-based systems (primarily electro-optical satellites in GEO and phased array radar satellites in LEO) and hundreds of HALE UAVs (primarily SAR equipped) to effect 24-hour surveillance of Chinese forces. These systems relay their data through communications satellites and HALE UAVs (with communications packages).* SATCOMs relay the data directly to established ground stations, then on to Cheyenne Mountain. The HALE UAVs relay their data to ground stations in Japan for relay via fiber-optic cable through the US communications grid to Cheyenne Mountain. State-of-the-art computers fuse this data.** Russian liaison officers review the resulting information and distill what's necessary for transmittal to Russian units. The US provides liaison officers and the requisite hardware and software (e.g., computers, satellite links, receivers, crypto keys, display boards) to the Russians.***

Degrade Chinese Stealth Cruise Missile Accuracy

Because Chinese cruise missiles use GPS for updating en route guidance, JFACC directs jamming of the unencrypted civil signal (CA code) from GPS satellites within view of the area of operations.**** Chinese cruise missiles can still navigate to the target area using cheap INS technology, but

---

*This is a feature of current UAV design efforts. For example, "the Tier 2 UAV will carry a 30-inch antenna for satellite communications." *Aviation Week*, 16 May 1994, 20. Whether the receiving ground station is in theater or in the CONUS is irrelevant to data transmission requirements.

**Algorithms are continually adjusted by contract programmers.

***We're assuming US and Russian units exercised this connectivity prehostilities.

****Liegh Ann Klaus, "GPS Advancements Overshadow Growing Pains," *Defense Electronics*, February 1995, 13–15. Retain Y code for US military operations.

without GPS updates they aren't as accurate as planned. As a result, Chinese cruise missiles become more susceptible to local jamming and spoofing in the target area. Units dedicated to this task are deployed with US force elements and attached to Russian air defense units.

Attack China's Stealth Aircraft Force

China bases its entire stealth aircraft inventory within Chinese borders. Due to political restrictions on homeland attack, US platforms deliver only nonlethal weapons against the Chinese stealth bases. Aerospace forces (cruise missiles, bombers) deliver conventional EMP and MHD bursts over Chinese stealth aircraft bases. These bursts cripple the manned stealth fleet. Those aircraft not effectively disabled have their sortie generation degraded (due to disabling of support equipment). These bursts also disable some C^4I nodes. To further degrade operations, cruise missiles spray anti-POL microbes on these bases. Despite these attacks, China remains capable of generating 24 stealth fighter-bomber sorties per night (.33 sortie rate).

Long-range bombers and cruise missile carriers launch from Alaska, the CONUS, Guam, Thailand, Vietnam, and Russia. They complicate Chinese defenses by attacking from all clock positions.

Defend Against Chinese Stealth Attack

Understanding that Chinese stealth systems will reflect or emit signals intermittently during flight, the US deploys a mix of sensors to the AO. The US deploys long-wavelength radars to northern Japan.* SIGINT collectors operate from Irkutsk and Elmendorf. Within the Maritime Provinces, the US deploys high-power, short-range radars and lidars across likely attack corridors. Sensors point both forward and backward to take advantage of frequency shortfalls in nonfrontal aspects. Radar reflections from pulses emitted from one location can be

---

*Despite Japanese neutrality in this war, Tokyo would likely authorize use of its terri - tory for air defense radars, especially if the data collected by these systems went directly to the US, not Russia .

received in a second location, allowing wide separation between sensor and shooter. UAVs launched from austere airfields (e.g., highway strips) patrol likely launch and penetration areas. They carry different types of sensors (e.g., electro-optical, inverse synthetic aperture radar, or jet-engine modulation). All UAVs transmit data directly to communications satellites for relay to the JFACC at Cheyenne Mountain.

JFACC assesses this data for accuracy. Because sensors are "looking" for small returns, false targets proliferate. JFACC's computers distill and filter the returns to provide an accurate picture. Based on this accurate picture, JFACC transmits taskings and information to air defense units. Stealthy interceptors respond from Petropavlovsk and bases on Sakhalin and the Northern territories. They're equipped with multispectral seekers on their air-to-air missiles. Radar solutions remain possible; during air-to-air engagements, enemy fighters lose sufficient "stealthiness" to enable tracking solutions. Imaging sensors prove decisive against cruise missiles. In addition to air defense fighters, a limited number of air defense batteries protect key US forces. They also receive direct tasking from JFACC. Finally, Russian liaison officers relay the threat picture to Russian forces, along with information on US responses.

Defend Against Chinese Theater
Ballistic Missile Attack

Russian forces deploy their own BMD to defend their own bases. JFACC deploys US BMD to defend only US forces. Because the Chinese use mobile launchers for their ballistic missiles, air strikes on Chinese missile launch facilities are not productive. JFACC includes ballistic missile C^{4}I within the homeland target list for bombers and missiles carrying nonlethal weapons. Missile production and test facilities are also in this list. The JFACC positions an airborne laser aircraft over the Maritime Provinces, but it is shot down by a Chinese stealthy fighter. Thereafter, BMD relies on space-based IR sensors to detect launches (satellites dedicated to this mission are augmented by covert sensors on other satellites and by

HALE UAVs orbiting over likely launch areas). After launch detection, JFACC cues theater and terminal defenses. These defenses use kinetic kill interceptors, ground-based, high-powered lasers, and directed energy weapons for the intercepts. Emitting ground-based sensors in the terminal areas are made mobile in order to degrade preemption attacks.

Attack Chinese Strategic and Operational Information

In support of the Joint Force Information Component Commander's campaign, aerospace operations center on two missions: (1) destroy key nodes (such as collection platforms, relay networks, and fusion centers) in the Chinese information network; and (2) distort data in the Chinese information network by viral insertion and spoofing. These twin operations delay vital information to the enemy while making the enemy mistrust the remainder.

Missiles and bombers deliver nonlethal weapons against C^4I nodes within China. Targets include the defense ministry, component headquarters (down to corps/fleet/regiment levels), C^2 of collection platforms, the communications system between the national leadership and its internal security apparatus, logistic depots and systems, power grids, banking, TV/radio, and the transportation network (especially the air traffic control system). JFACC also orbits high-altitude UAVs overhead Chinese satellite downlink facilities. Equipped with jammers, these UAVs disrupt the tasking and data flow between ground stations and satellites.

Within Russia, the JFACC directs more conventional destructive means against Chinese information systems. Targets include theater C^2 facilities, collection platforms, transmitters, transmission lines, logistics, and information repair facilities/units. Methods include precision bombs, jamming, and spoofing. Because the goal is to dominate information at the strategic and operational levels of war, the JFICC orchestrates these attacks. The JFACC only executes its aerospace aspects.

Deny Chinese Invasion Forces Mass, Maneuver, and Effective C^{4}I

Cruise missiles launch from ground, sea, and air platforms. Sea-launched cruise missiles launch from beyond 1,000 NM due to the "counterbattery" threat of Chinese cruise missiles .* Air-launched cruise missiles (ALCM) launch >500 NM from China due to the stealth fighter threat.** Ground-launched cruise missiles launch from Russian bases in the Kurile Islands. Prevailing winds in the area aid the US attack. Winds are generally out of the east at 15 knots.***

Long-range stealthy bombers launch from CONUS.**** They're escorted over the target area by long-range stealth fighters and UAVs (for jamming and suppression). After strike, the bombers recover at Guam, Elmendorf, and/or Irkutsk. Bombers return the next night and restrike Chinese targets en route back to CONUS. Air refuelers launch/recover in Russia, Alaska, and Taiwan. Fighters launch/recover from Petropavlovsk and Irkutsk.

The immediate task of this campaign is to drop all major bridges across the Ussuri and Amur Rivers on the Sino-Russian border east of 130E. The intent is to cut supply lines in Russia and China. One stealth bomber sortie effects these cuts the first night of the war. Resulting choke points are identified by satellites and UAVs (with wide-area automatic target recognition [ATR]), then attacked by cruise missiles during the day and long-range stealth bombers at night.

Stealthy cruise missiles and bombers deliver a wide variety of munitions on Chinese forces. Scatterable mines, anti-POL gels, and conventional EMP blasts break unit cohesion and canalize the enemy advance. Brilliant munitions destroy high-value equipment. Each sortie results in multiple kills.

---

*Submarine-launched cruise missiles have little utility in this scenario.

**Optimum cruise missile carrier is an "ALCM truck." It would have long-range and large weapons capacity.

***850-millibar and 700-millibar vector mean winds are 080/10 and 090/10, respec - tively. Data furnished by USAFETAC/DOO.

****0.5 sortie rate due to CONUS basing. Quick turns made possible by high (4.0) crew ratios. Assuming advances in explosives and precision over the next 15 years, each B-2 carries 40 weapons.

When Chinese forces disperse to offset these attacks, their assaults on Russian positions break down.

Against targets with limited windows of vulnerability (sometimes called "short dwell" targets), JFACC launches ballistic missiles. Strikes are highly successful due to sensor-to-warhead target data transmission, the NRT decision cycle, and maneuvering warheads capable of identifying/tracking mobile targets (e.g., mobile C^2, Chinese cruise missile TELs on Russian territory). Chinese ballistic missile defenses are ineffective in the forward area.*

Airlift Critical Supplies to Russian Forces

The US limits airlift deliveries to high-value replacement parts/weapons. SAM reloads top the priority lists, followed by UAVs and air-to-air missiles. Due to the Chinese cruise missile threat, off-loads are rapid. Airlift sorties landing in the Maritime Provinces spend less than one hour on the ground. Their goal is to arrive and leave before Chinese C^4I can detect the airlifters' presence, direct an attack, and deliver warheads into the target area.**

Prosecute a Parallel War

The US attacks key nodes of Chinese military power across all levels of war with near-simultaneity. As stated before, the NCA restricts attacks on the Chinese political structure (due to the threat of nuclear retaliation). All other targets, however, are fair game for attack. At the tactical level, the idea is to defeat the attacking invasion force—not just by attacking any single system (such as armor), but by concurrent attacks on all facets of the invasion force. This includes the armor, logistics (e.g., POL, ammunition), personnel, command and control, and LOCs (especially roads, tunnels, and bridges).

At the operational level, the idea is to defeat the attacking armies—not just by attacking individual units, but by

---

*Just as today's mobile SAMs in forward areas are less effective than fixed SAMs around capital cities.

**Time-of-flight assumption based on a subsonic cruise missile launched 500 NM away.

concurrent attacks on units, their theater supplies, their supporting personnel, the command and control links between field commanders and units, and their invasion LOCs, especially the bridges across the Amur and Ussuri Rivers.

At the strategic level, the idea is to defeat China's ability to wage war—not just by severing communications between the national leadership and its elements of power, but by concurrent attacks on the national leadership's communication infrastructure, China's national communications and transportation grids (with nonlethal weapons), and its air defenses.

All these attacks (strategic, operational, and tactical) occur *near simultaneously*. Such simultaneity is a daunting challenge. Given current technology, parallel war against a peer enemy is not possible today. However, given current *trends* in technology, this capability could exist in 2010. Improvements in information will allow greater efficiency in delivery, thus lessening wastage. Decreasing target location error and munition delivery error will substantially decrease the size and number of bombs needed for a high Pk. The end result will be thousands of munitions delivered precisely against critical targets in a highly compressed time period. For example, it is reasonable to project a future force of 100 bombers, each carrying 800 precision munitions. That's a total of 80,000 bombs delivered precisely against specific targets. Several thousand cruise missiles with a wide variety of munitions (e.g., EMP, HPM, MHD, antiradiation) would precede the bombers, hitting air defense radars (particularly low frequency, antistealth radars), antenna arrays, jammers, electrical grids, and soft C^4I. This assault, repeated daily for weeks, would shock the Chinese offensive, enabling Russian forces to hold their ground.

Appendix B

Illustrative Scenario (Niche)

To illustrate a future war with a niche competitor, imagine the following hypothetical scenario. Nigeria attacks Zaire with 10 armored divisions (including 5,000 armored vehicles). Nigeria's operational goal is to capture Kinshasa, thereby cutting Zaire's access to the Atlantic.

Drawing on its oil wealth, Nigeria has built a substantial military. In addition to the ground attacking force, Nigeria's military consists of:*

- 200 fighter aircraft (half are front line: MiG-29, Rafael, Su-35), equipped with advanced medium-range air-to-air missiles (AMRAAM) and precision ground-attack munitions,
- Five nuclear weapons (100 kilotons), and
- 100 intermediate-range ballistic missiles (IRBM) with a range of 1,000 NM and CEP of more than 500 meters.

Also, Nigeria has integrated three emerging weapons into its military. First, it uses COMSATs (commercial satellites) for photo reconnaissance and communications. These satellites are majority-owned by the European Community, the Arab League, and an Asian business consortium. Nigeria is a minority owner of each system. All refused US requests to shut down Nigerian access. These satellites allow encrypted communications throughout Africa. They also provide imagery directly to mobile ground stations in Nigeria and with the attacking columns. This imagery has one-meter resolution. In addition, the Nigerians have two backup ground stations in sympathetic third countries. These facilities are connected to Nigeria by fiber-optic cable.

Second, Nigeria has stealthy cruise missiles (purchased from Israel, Brazil, and France) with these capabilities:

- Inventory: 5,000, with 100 mobile launchers.
- Guidance: Integrated GPS/INS en route; terminal guidance uses millimeter wave seeker, infrared, or antiradiation. CEP is <5 meters.

---

*Such a force is notional; it bears no relation to current Nigerian inventories. In total numbers it's similar to the force structure used by RAND in *The New Calcu lus* 16.

135

- Range: 1,000 NM.
- Warhead: 1,000-pound conventional warhead, cluster bomb unit (CBU), zirconium disk, napalm, infrared cameras with data burst transmitters, multichannel jammers, decoys, HPM, MHD, or EMP.
- RCS <.001 (frontal aspect only).

Third, Nigeria bought five ASATs from Russia. The ASATs launch from Su-35s. They are kinetic kill with IR sensors. These ASATs can only reach satellites in low earth orbit.

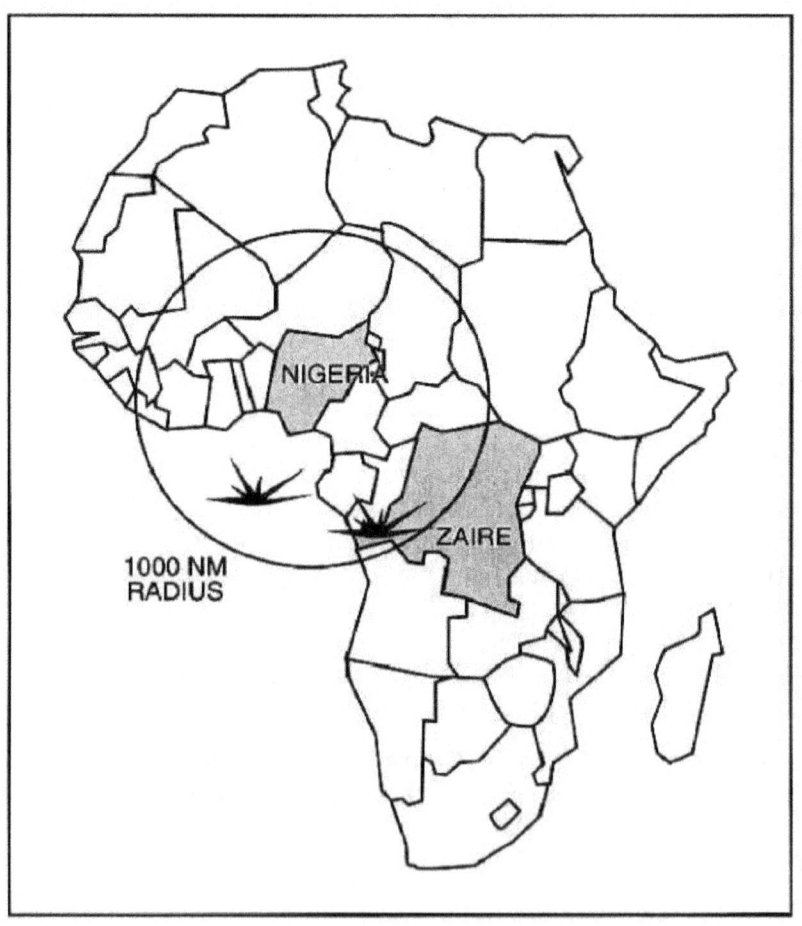

Figure 10. Map of Africa

Zairian forces are moving to reinforce Kinshasa but are considered far inferior to the Nigerian military. Countries between Nigeria and Zaire (Cameroon, Gabon, and Congo) either ally with Nigeria or give Nigerian forces free passage. The US decides to unilaterally enter the conflict on Zaire's side.* The president orders the US military to assist Zaire's army in halting the Nigerian advance north of the Congo River.

The theater CINC levies the following specific aerospace taskings:

- Establish air supremacy over US forces.
- Establish air superiority over Zaire and the Nigerian LOCs.
- Airlift US units to the theater.
- Attack Nigerian military forces south of the Nigerian border.
- Provide near-real-time information on Nigerian military movements to Zaire.

Assuming this war starts in summer, weather is not much of a factor. Throughout the year, ceilings are above 3,000 feet more than half the time. The rainy season around Kinshasa is early winter, especially November and December. Precipitation in July and August is practically zero. Temperatures are always above freezing.

Table 6

Kinshasa Weather (Ceilings in Percent)

	Jan	Feb	Mar	Apr	May	Jun	Jul	Aug	Sep	Oct	Now	Dec
Ceilings <3,000'	22	19	19	19	25	49	40	45	33	30	16	16
Ceilings <1,000'	1	1	1	1	1	1	-	-	-	-	-	-
Inches Precipitation	5.0	5.6	6.7	8.4	5.4	.2	.1	.1	1.3	5.5	9.3	6.7

Source: USAFETC/DOO (Now AFCCC)

---

*This is only an illustrative scenario. The specific reason for intervening is not impor-tant for scenario purposes. What *is* important is to outline a contingency where the niche enemy is superior to its regional opponent. Also, the fight is thousands of miles from US territory and there's little in-place US military infrastructure.

The US begins deploying two armored divisions and 10 fighter squadrons, with their associated C^{4}I, into Zaire and southwestern Congo. One ARG and two carrier battle groups position themselves off the coast of Gabon. Defensive systems against cruise and ballistic missiles deploy with these forces. Batteries position around the land bases. Improved AEGIS-type ships escort the battle groups. Both rely on airborne aircraft equipped with long-wavelength radars for missile detection and cueing. Surface radars guide the actual intercepts.

One week after deployment begins, Nigeria launches 200 stealthy cruise missiles with CBU (plus four with electronic decoy packages and four with jammers) against the Brazzaville and Kinshasa airports. They overwhelm the limited number of fully operational air defense batteries.* In addition, some of the intercepts are hampered by software shortfalls; the Nigerian cruise missiles present slightly different signatures from those expected. The simultaneous arrival of three tactical ballistic missiles further confuses the defenses. As a result, half the cruise missiles penetrate. Multiple AWACS, J-STARS, KC-10s, and B-747s are destroyed on the ramp. A large tent city is heavily damaged. US casualties are high. Television crews transmit pictures of Zairian soldiers looting the bodies of dead Americans. Stealthy cruise missiles also cripple three USN ships off the West African coast and one RO/RO supply ship at Kinshasa harbor.

In the wake of this disaster, public support for intervention polarizes in the US. The president reaffirms the US commitment, but replaces the theater CINC and requests a new strategy. Planning is constrained by basing restrictions. All countries within 1,000 NM of the AOR refuse US overflight and basing requests. The US must rely on Ascension Island (United Kingdom), Azores (Portugal), Canary Islands (Spain), South Africa, Kenya, and Zaire for land bases. Bases in the CONUS, Europe, South America, and at Diego Garcia are also available.

*For example, a single Patriot battalion is normally equipped with six batteries, each of which can fire up to eight missiles before reloading (total: 48 missiles).

US Aerospace CONOPS

In this scenario, a hypothetical enemy (Nigeria) launches a land invasion of a nation 700 NM to its south (Zaire). This operation is a land force-intensive attack, with 10 divisions of armor and infantry.* Its size and capability far outmatch the defending (Zairian) army. To protect US interests in the region, the president orders the US military to assist Zaire. The military task is to stop the Nigerian advance short of the Congo River.

Anticipating American intervention, Nigeria bought high-technology weapons. It has stealthy cruise missiles with 1,000 NM range, access to civilian satellite information and communications, and a robust air defense system (radars, fusion centers, surface-to-air missiles, and manned interceptors).

In accordance with guidance from the theater commander, the US aerospace CONOPS contains a baker's dozen of major themes.

· Limit the number of targets vulnerable to attack.
• Command US aerospace operations from Cheyenne Mountain.
  • Protect US space systems.
  • Collect NRT information on enemy maneuvers.
  • Degrade Nigerian stealth cruise missile accuracy.
  • Attack strategic and operational targets within Nigeria.
  • Defend against Nigerian stealth attack.
  • Defend against Nigerian ballistic missile attack.
  • Attack Nigerian information at all levels of war.
  • Attack Nigerian ground forces.
  • Airlift critical supplies to Zairian forces.
  • Support the ground counteroffensive.
  • Prosecute a parallel war.

Limit the Number of Targets Vulnerable to Attack

Absent a direct threat to US vital interests, it is doubtful the American public will support a casualty-intensive war (e.g., Vietnam, Korea). Therefore, casualties are a US strategic center

---

*Current demographic projections estimate Nigeria's 2010 population at 240 million.

of gravity. US forces must accomplish their mission while (1) minimizing the number of US forces deployed within range of Nigerian stealthy cruise missiles, and (2) protecting those forces deployed to the area.

Accordingly, most US aerospace forces base more than 1,000 NM from Nigeria, outside the range of Nigeria's cruise missiles. They launch from bases in the CONUS, Europe, islands in the Atlantic (Azores, Canaries, Ascension), Brazil, South Africa, southern Zaire, Kenya, and Diego Garcia. This basing structure puts a premium on long-range bombers and missiles. Short-legged, nonstealthy aircraft (e.g., F-16, F-18) have little use.

To support Zairian forces and to coordinate actions, a limited number of US forces deploy near Kinshasa. These forces provide "enhancing" capabilities: C^4I, air defense (including spoofing and jamming teams), surface-to-surface missile batteries (e.g., cruise missiles), UAV launchers, and airlift infrastructure. One squadron of close-air-support helicopters also deploys for emergency assistance.

These forces disperse after landing. They remain mobile, shifting location every day or two. They have robust aerospace defenses. Where possible they use Phalanx for point defense and reactive armor to decrease warhead explosive effect. For sustenance they require only a few airlift sorties a day.

Command US Aerospace Operations from Cheyenne Mountain

Given (1) the dispersed nature of US forces, (2) the need to minimize US exposure to casualties, and (3) the lack of a need to coordinate aerospace actions with other air forces, the JFACC remains in the CONUS.* From a secure, well-exercised base the JFACC commands the worldwide attack on Nigeria and its invasion/occupation forces.

---

*Cheyenne Mountain has the advantage of proximity to USSPACECOM. Other candi -date locations could include Falcon AFB (also has excellent connectivity with USSPACECOM); Langley AFB (wealth of personnel with operational expertise); and the Pentagon (ready access to satellite downlinks, intelligence, political leadership). In all cases, significant parts of the JFACC's staff could be located elsewhere. For example, logistics could be coordinated from Langley AFB while CHECKMATE in the Pentagon drafts strategic targeting.

JFACC HQ receives all sensor data from satellites and UAVs via SATCOM relay. JFACC tasks strike aircraft in the CONUS (e.g., stealth bombers at Whiteman AFB), strike aircraft in theater (e.g., FB-22s at Nairobi), air defense and battle management units in theater (based in southern Zaire), cruise missile-equipped ships at sea, and air refuelers at Ascension Island, Tenerife (Canary Islands), and Lajes Field (Azores). JFACC HQ works closely with USSPACECOM to manage and position reconnaissance satellites, plus deconflict communications satellite requirements.

As a backup to SATCOM links, the Joint Force Information Component Commander (JFICC) deploys HALE UAVs with communications relay packages to the AOR. A bridge of three HALE UAVs connects theater assets with a downlink station on Ascension Island. Fiber-optic cable moves this data from Ascension Island to the US information grid (through Europe). However, due to successful attacks on Nigeria's space-tracking facilities, this alternate routing is unnecessary.

ATOs contain taskings, support data, and intelligence information. A central intelligence management staff attaches relevant information to ATO taskings. Units retain the authority to query the database for more information, but the normal procedure—facilitated by computer software—includes applicable intelligence with each tasking.

The theater CINC deploys forward. The CINC gives general guidance (e.g., apportionment, overall objectives) through a liaison office to the JFACC. The JFACC puts staff elements forward in Kinshasa to liaise with Zairian forces (especially army units and ADA).

Protect US Space Systems

Nigeria's nuclear weapons (i.e., EMP) and ASATs threaten US space systems, particularly those in LEO. To preclude an EMP burst in space, US cruise missiles attack Nigerian space launch facilities on day one of hostilities. Space launch facilities are fragile. They do not stand up to multiple explosions.

To deal with the air-launched ASAT threat, the US takes three approaches. First, JFACC gives a high priority to attacks against those airfields where ASAT-capable aircraft (i.e.,

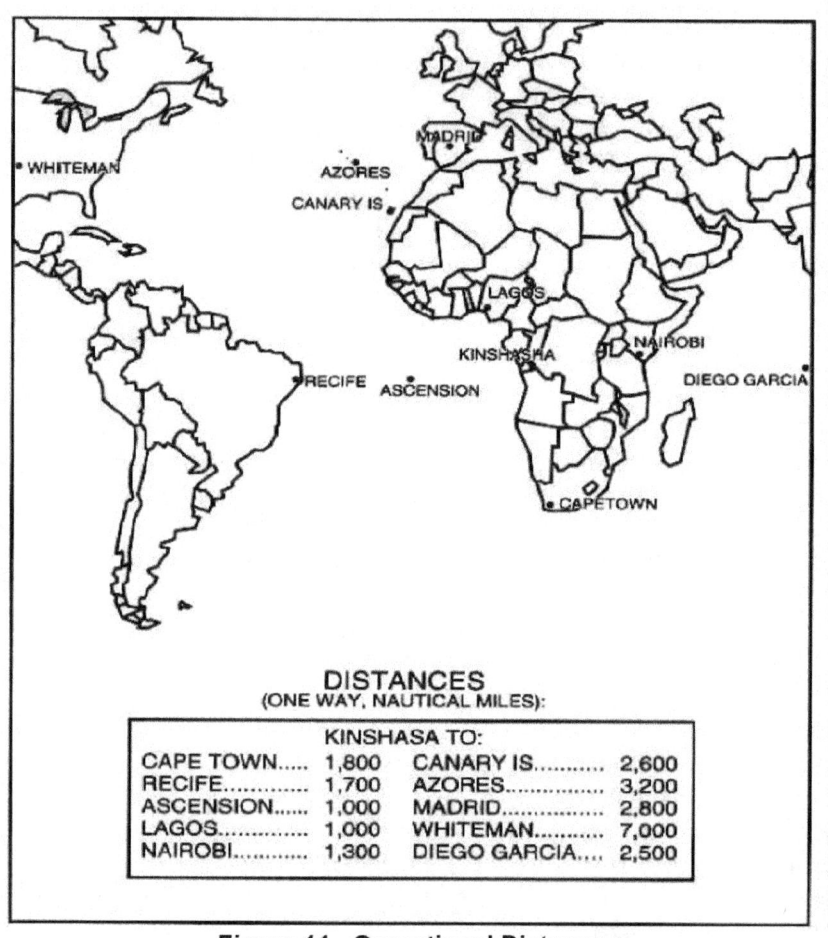

Figure 11. Operational Distances

SU-35) are based. Both lethal and nonlethal weapons go against aircraft shelters, storage facilities, C^4I, barracks, and power sources. Despite this effort, some ASATs survive. As a second approach, the US readies replacement platforms (satellites and UAVs). Because the niche enemy has a small, finite ASAT capability, the US does not need a robust space protection regime (shielding, maneuverability, stealth, escort satellites, etc.), as would be required for war with a peer

142

competitor. In this case, relying on replacement satellites is cheaper than incorporating robust defenses on each satellite. Third, the US destroys Nigeria's space tracking facility. Without the means to observe normal orbit variances, Nigeria's ASATs become too inaccurate to be militarily effective.

Collect NRT Information on Enemy Maneuvers

High-value airborne platforms (AWACS, J-STARS, TR-1, Rivet Joint) launch and recover from airfields in southern Zaire (Katanga Province). Bases include Kolwesi, Likasi, and Labumbashi. SIGINT platforms operate from Ascension Island. Co-based fighters escort surveillance platforms, although orbits remain outside the combat radius of Nigerian fighters (unrefueled).

J-STARS proves exceptionally worthwhile in this scenario. From orbits 150 NM either side of the Nigerian advance (outside of SAM range), J-STARS maps advancing columns. It identifies tracked versus wheeled vehicles. It provides adequate targeting information to long-range strikers, and identifies logistics sites through history traces of supply vehicles. No substantial forces move without detection by J-STARS.

Although their orbits are within range of Nigeria's cruise missiles, and these cruise missiles have antiradiation sensors, Nigeria's cruise missiles do *not* pose a threat to emitting surveillance aircraft. The AWACS has a limited ability to detect stealthy cruise missiles within a few miles of its antenna. Once the AWACS detects a cruise missile inbound against it, the AWACS causes the antiradiation sensor on the Nigerian cruise missiles to break lock by "blinking" its radar and changing its flight path. Unlike the missiles of a peer competitor, Nigeria's missiles have only a single air-to-air sensor (antiradiation). Manipulation of the AWACS signal causes the cruise missile to break lock. Other emitting surveillance aircraft (such as J-STARS) orbit close enough to the AWACS to receive similar warning information. They also use blink and maneuver to defeat antiradiation sensors. Finally, both the AWACS and J-STARS have active defenses against missiles (e.g., chaff, flares, EW).

HALE UAVs launch from Kinshasa to patrol Nigerian maneuver forces, especially those beyond J-STARS range. For the most part, UAVs patrol 100 NM to the east and west of Nigerian ground forces. These UAVs fly parallel to the Nigerian advance but outside the range of Nigerian SAMs. Data collected from all UAVs are uplinked to communications satellites for relay to the CONUS JFACC.

Sensor suites include beam radar, EO, phased-array radar (with MTI), and IR. The beam radar is the lightest and cheapest. It sweeps likely areas with its radar beam. It filters returns through a wide-area ATR package on the UAV looking for certain patterns. When it identifies these patterns, it transmits suspected target locations and types to the JFACC. The JFACC then tasks UAVs carrying more defined sensors (EO, SAR, and IR). These sensors can discriminate between tanks and trucks with high confidence. They have sufficient resolution to identify vehicles by type (e.g., T-72 versus T-60). This dual UAV approach allows wide-area coverage with only a few high-cost EO, phased-array radars, and infrared-capable UAVs.

The JFACC orchestrates atmospheric systems and satellites to effect 24-hour conflict surveillance. Revisit times are 30 minutes or less. State-of-the-art computers fuse this data. Zairian liaison officers review the resulting information and distill what's necessary for transmittal to Zairian units. The US furnishes the transmission means (e.g., computers, satellite links, receivers) plus liaison officers to Zairian forces.

Degrade Nigerian Stealthy Cruise Missile Accuracy

Nigeria's cruise missiles use GPS to update their en route navigation system. The US degrades Nigeria's cruise missiles by manipulating GPS civil mode within 1,000 NM of Nigeria. For political reasons, turning off the civil code (course acquisition [CA] code) on those satellites within the field of view of the route/target area is *not* an option. Civil authorities throughout Africa require GPS for everyday functions. Signal manipulation is through a combination of local jamming and spoofing. Ten- to 25-watt jammers deploy around and among US forces and key target areas. Jamming is accomplished without affecting the military GPS signal (Y code) that is needed for US operations.

Degrading cruise missile accuracy is *not* an "additional duty" for aerospace defense units. It is performed by units specifically structured, trained, and dedicated to this job. They direct intelligence services to identify weaknesses in Nigerian cruise missile hardware and software. Nigerian cruise missiles' target recognition software is a prime essential element of information (EEI). By disturbing the expected pattern electronically and by positioning decoys, air defense units degrade the accuracy of Nigeria's cruise missiles.

Attack Strategic and Operational Targets within Nigeria

The JFACC orchestrates an aerospace campaign against weapons stores, airfields, infrastructure (including civil electrical and communications grids) and strategic C^{4}I targets in Nigeria. Primary means are long-range stealthy bombers and long-range cruise missiles.

Cruise missiles launch from surface ships and submarines. The US Navy deploys with 1,000 stealthy cruise missiles, each with a range of 1,000 NM. Their warheads include 1,000-lb conventional warheads, CBU, and EMP. They have brilliant sensors (with ATR) and inertial guidance. They operate 24 hours a day.

Stealth bombers launch and recover in Recife, Brazil.* Aircraft fly a 0.5 sortie rate carrying 16 PGMs per sortie.** Their air refueling support comes from Ascension Island. Bombers are the primary platform for attacking hard targets. They operate over the target area only at night, escorted by stealth fighters and UAVs (for jamming and suppression).

---

*This profile assumes an unprogrammed deployment capability on the part of the B-2. If no such capability exists, the B-2s would have to sortie from the CONUS, which would cut their sortie generation in half. Operating from the CONUS might also degrade operational responsiveness because the longer flight times would require ear - lier aircrew mission planning, which would require an earlier posting of the ATO. This profile also requires a high (4.0) crew ratio.
**A force of 20 B-2s would result in 160 precision bombs per day (20 x .5 x 16 = 160). A force of 40 B-2s would equate to 320 precision bomb deliveries per day.

Defend Against Nigerian Stealth Attack

Air defenses use multiple types and layers of area sensors to detect enemy stealth, then fuse this information to allow rapid cueing and focused targeting. Understanding that stealth systems reflect or emit signals intermittently during flight, the US deploys a mix of sensors to the AOR. Long-wavelength radars deploy to Kinshasa. Within Zaire and the Congo, the US deploys high-power, short-range radars and lidars across likely attack corridors. They're arranged in a circular, overlapping fashion to detect stealthy cruise missiles from the side and rear aspects. Radar reflections from pulses emitted from one location can be received in a second location, allowing wide dispersion between sensors and shooters. Finally, UAVs launched from austere airfields (e.g., highway strips) patrol likely launch and penetration areas with a mix of sensors. All transmit data to SATCOMs for relay to the JFACC in the CONUS.

JFACC fuses this data and retransmits taskings and information to air defense units. Stealth interceptors, flying in a CAP role, respond. The interceptors are equipped with multispectral seekers on their air-to-air missiles. Radar solutions remain possible at short range. Imaging sensors prove decisive against cruise missiles. In addition to air defense fighters, a limited number of air-defense batteries protect key US forces. They also receive direct tasking from JFACC. Finally, Zairian liaison officers relay the threat picture to Zaire's forces, along with information on US responses.

Defend Against Nigerian Ballistic Missile Attack

Nigeria has 100 IRBMs with 1,000-NM range. Because they're launched from mobile TELs, attacks on launch sites are impractical. Instead, the JFACC directs a five-stage defense:

(1) Bombers and cruise missiles attack Nigeria's ballistic missile C^4I. These attacks degrade Nigeria's ability to employ its missiles to maximum effect. (2) Space-based systems (e.g., DSP, SBIR) detect ballistic missile launches, then cue radar trackers. (3) Airborne and ground radars use this cueing to

146

detect and track the missile. (4) Intercepts are accomplished by airborne lasers and ground-based missiles. (5) Because Nigeria's ballistic missiles are equipped with maneuvering warheads, manipulating the GPS and the use of spoofing signatures in target areas lessens warhead accuracy. Despite these efforts, one in 10 missiles penetrates. Of the successful penetrators, half hit valuable targets.

Attack Nigerian Information at all Levels of War

Because the goal of these attacks is to dominate information at the strategic and operational levels of war, the JFICC orchestrates these attacks; the JFACC executes only its *aerospace* aspects.

Aerospace operations center on two areas. One is destroying nodes (such as collection platforms, relay networks, and fusion centers) in the Nigerian information network. The other is distorting information in the Nigerian information network by viral insertion and spoofing.

Missiles and bombers attack all aspects of Nigeria's information system. This includes political, economic, and military information systems. Targets include the information aspects of leadership, production, power, transportation, public information, and the military. Aerospace forces also target all UAVs and satellites the niche owns.

In most cases, targets require precision. Targets include connectivity (e.g., fiber-optic lines and radio/cellular antennas), nodes (e.g., switching stations), repair assets, downlink stations (e.g., satellite dishes), fusion centers, platforms (aircraft, UAVs, and satellites), and C^2 personnel. Munitions cover the gamut of the inventory: earth penetrators, MHD, EMP, HPM, CBU, HE, and so forth. Bombers and cruise missiles deliver the munitions.

Attack Nigerian Ground Forces

The goal is to "hollow-out" the Nigerian advance to a point where indigenous defenses can hold. This point can be accomplished by denying the Nigerians mass, maneuver, and effective C^4I.

The Nigerians must move 10 divisions over 700 miles, carrying all their logistics with them. Due to the distance from Nigeria, these forces move without air cover. The terrain is difficult. The area between Nigeria and Zaire is mountainous and laced by major rivers. Choke points abound.

US bombers deliver high-explosive, penetrating bombs (e.g., I-2000) against bridges and tunnels in advance, behind, and between the Nigerian divisions. They scatter mines and unattended ground sensors (UGS) along likely routes of advance. Bombers then hit the resulting tie-ups with a mix of munitions: sensor fused weapons, WAM, BAT, and CBU. Trucks (especially tank transporters, fuel carriers, and food wagons) are the highest priority.

US bombers launch from Nairobi and Mombasa, Kenya, hit their targets (armored columns, LOCs, logistics), and return.* They fly a 2.0 sortie rate.** Aerial refuelings are not required.

Fighter-bombers based in southern Zaire escort these strikes over the target area. Escorts carry a mix of defense suppression missiles, air-to-air missiles, and precision bombs. These fighter-bombers need long legs. Due to minimal airport infrastructure in southern Zaire, the JFACC would prefer not to base large numbers of tankers at these bases. Tankers take up ramp space and drain limited jet-fuel supplies.

Note: Nigerian forces routinely carry low-power jammers tuned to the GPS frequency. The jammers degrade US systems that are dependent on GPS in the target area (e.g., munitions such as JDAM).

Airlift Critical Supplies to Zairian Forces

Once Nigerian C^2 is stressed and its NRT surveillance destroyed, airlift operations into Kinshasa and Brazzaville resume. Because Nigeria retains its stealthy cruise missiles and access to civilian overhead, airlifters must vary arrivals and limit ground times to only a few hours.

---

*Requires overflight approval by *one* of the following: Sudan, Uganda, or Tanzania.
**Quick turns made possible by high (4.0) crew ratios. Due to minimal infrastructure in Kenya, spare aircraft and attrition reserve would base at Diego Garcia.

148

Airlifters fly either directly from the CONUS or stage from Lajes (Azores). Their air refueling support comes from Lajes, Tenerife, and Las Palmas (Canary Islands). Because these airports have limited fuel supplies, the US offers to pay "spot plus 50 percent" to private firms to keep each base flush with jet fuel.

Support the Ground Counteroffensive

Aerospace forces support the ground component commander's (GCC) campaign. The CINC's guidance is to drive advanced elements of Nigeria's army away from Zaire to at least the Cameroon-Congo border. In accordance with the GCC's plan of maneuver, aerospace forces provide heavy firepower support while continuing to interdict Nigerian army elements.

Although an amphibious landing in Cameroon has the potential for cutting off the Nigerian army from its home base, Nigeria's remaining cruise missile inventory and access to civilian overhead imagery dissuades the CINC from this option.

Prosecute a Parallel War

The US attacks key nodes of Nigerian power across all levels of war nearly simultaneously. Operational and strategic attacks offer the most leverage.

At the operational level, the idea is to defeat the attacking 10 divisions—not just by attacking individual units—but by concurrent attacks on their invasion LOCs (especially the tunnels and bridges in Cameroon, Gabon, and Congo), their C^2 (the mechanism for directing repairs), and logistics stocks of fuel and food. When these target sets are attacked at the same time, the attacking divisions have too much to fix, too few resources to fix it, and no way to efficiently direct what's left.

At the strategic level, the idea is to defeat Nigeria's ability to wage war. Due to its possession of nuclear weapons, the NCA does not permit decapitation attacks on Nigerian leadership. However, the NCA permits attacks intended to *isolate* the Nigerian leadership from its elements of power. These attacks sever normal communications between the national leadership and its elements of power. Simultaneous attacks also degrade

149

the leadership's access to near-real-time information, strip homeland air defenses, effectively stop all centralized electrical production, cut the national transportation grid, and place the populace at risk—all in a short period of time. As a result of these simultaneous attacks across all strategic centers of gravity, the Nigerian leadership finds itself unable to orchestrate an effective war with Zaire.

At the tactical level of war (where individual attacks actually occur), aerospace forces attack all elements of a unit's strength nearly simultaneously. In addition to attacking individual weapons (e.g., a tank, airplane, or ship), aerospace forces attack the C^2 for that weapon, its communications, intelligence support, defenses, morale, and supplies. Attacks on the weapon *system*, as opposed to just the weapon, within a short time period reduces that weapon's ability to adapt.

This aerospace campaign, across levels of war and all target sets, would destroy the Nigerian offensive.

Appendix C

List of Acronyms

AADC	Area Air Defense Commander
ABCCC	Airborne Battlefield Command and Control Center
ABL	airborne laser
ABM	antiballistic missile
ACA	airspace control authority
ADA	air defense artillery
AEGIS	airborne early warning/ground environment integration segment
AFCCC	Air Force Combat Climatology Center
AIFV	armored infantry fighting vehicle
AIM	air intercept missile
ALCM	air-launched cruise missile
AMRAAM	advanced medium-range air-to-air missile
AOC	air operations center
AOR	area of responsibility
APC	armored personnel carrier
ARG	Amphibious Ready Group
ASAT	antisatellite weapon
ASOC	Air Support Operations Center
ATACMS	Army Tactical Missile System
ATO	air tasking order
ATR	automatic target recognition
AWACS	Airborne Warning and Control System
BAM	Baikal-Amur Mainline Railroad
BAT	brilliant antitank munition
BEF	British Expeditionary Force
BDA	battle damage assessment
BMD	ballistic missile defense
BUR	bottom-up review
BVR	beyond visual range
C^2	command and control
C^3I	command, control, communications, and intelligence

153

C⁴I	command, control, communications, computers, and intelligence
CA	course acquisition
CAP	combat air patrol
CBU	cluster bomb unit
CCD	camouflage, concealment, and deception
CDC	concepts development center
CENTCOM	US Central Command
CEP	circular error of probability
CIA	Central Intelligence Agency
CIC	Combat Information Center
CINC	commander in chief
CINCSPACE	Commander in Chief, Space Command
CIO	Central Imagery Office
CJCS	Chairman, Joint Chiefs of Staff
COMSAT	commercial satellite (civilian-owned)
CONOPS	concept of operations
CONUS	continental United States
CRAF	civil reserve air fleet (MAC)
CVBG	Carrier Battle Group
CVN	aircraft carrier (nuclear powered)
D-ASAT	defensive antisatellite weapon
DEW	directed-energy weapons
DIA	Defense Intelligence Agency
DISA	Defense Information Services Agency
DMA	Defense Mapping Agency
DPRK	Democratic Peoples Republic of Korea
DSP	Defense Support Program
EEI	essential elements of information
EHF	extremely high frequency
ELINT	electronic intelligence
EMP	electromagnetic pulse (<MHZ)
EO	electro-optical
EUCOM	European Command
EW	electronic warfare

FAA	Federal Aviation Administration
FYDP	Future Years Defense Plan
GCC	Ground Component Commander
GEO	geosynchronous earth orbit (22,300 miles)
GPS	global positioning system
HALE	high altitude, long endurance
HE	high explosive
HEO	highly elliptical orbit
HPM	high-powered microwave (100-200MHZ)
HTACC	Hardened Tactical Air Control Center
HUMINT	human intelligence
ICBM	intercontinental ballistic missile
IMINT	imagery intelligence
INF	intermediate-range nuclear force
INS	inertial navigation system
IR	infrared
IRBM	intermediate-range ballistic missile
ISR	intelligence, surveillance, reconnaissance
IW	information war
J-3	Operations Directorate (Joint)
J-5	Strategic Plans and Policy Directorate (Joint)
J-6	Command, Control, and Communications Systems Directorate (Joint)
JAOC	Joint Air Operations Center
JCS	Joint Chiefs of Staff
JDAM	Joint Direct Attack Munition
JFACC	Joint Force Air Component Commander
JFICC	Joint Force Information Component Commander
JTCB	Joint Targeting Coordination Board
J-STARS	joint surveillance target attack radar system
KT	kiloton
KTO	Kuwaiti theater of operations

LEO	low earth orbit (60-250 miles)
LF	low frequency
LGB	laser-guided bomb
LIDAR	light detection and ranging (a laser radar)
LO	low observable (~ -10 to -15 db)
LOC	line of communications
LOS	line of sight
LRC	lesser regional contingency
LRR	long-range radar
MASINT	measure and signature intelligence
MBT	main battle tank
METT-T	mission, enemy, terrain, troops-time
MHD	magneto-hydrodynamic (1-100MHZ)
MILSTAR	military strategic and tactical relay satellite
MRC	major regional contingency
MTI	moving target indicator
NATO	North Atlantic Treaty Organization
NBC	nuclear, biological and chemical
NCA	National Command Authorities
NLOS	non-line-of-sight
NM	nautical miles
NRO	National Reconnaissance Office
NRT	near real time
NSA	National Security Agency
OODA	observe-orient-decide-act
OPCON	operational control
OPS	operations
PACOM	Pacific Command
PGM	precision guided munitions
Pk	probability of kill
PLA	Peoples Liberation Army (China)
PLAAF	Peoples Liberation Army Air Force (China)
POL	petroleum, oil, and lubricants
POM	program objectives memorandum
PSYOPS	psychological operations

RCS	radar cross section
R&D	research and development
RMA	revolution in military affairs
RO/RO	roll-on/roll-off (USN)
ROCC	Regional Operations Control Centers (NORAD)
ROE	rules of engagement
SAM	surface-to-air missile
SAR	synthetic aperture radar
SATCOM	communications satellite (military)
SBIR	space-based infrared system
SBWAS	space-based wide-area surveillance
SFW	sensor fused weapon
SIGINT	signals intelligence
SIOP	Single Integrated Operations Plan
SOF	Special Operations Forces
SOSUS	sound underwater surveillance system
SSN	attack submarine (nuclear powered)
TACC	Tactical Air Control Center
TBM	tactical ballistic missile
TCT	time-critical target
TEL	transporter-erector-launcher
TLAM	Tomahawk land-attack missile
TR-1	tactical reconnaissance aircraft (U-2 derivative)
UAV	unmanned aerial vehicle
UGS	unattended ground sensors
UHF	ultra high frequency
USA	United States Army
USAF	United States Air Force
USAFETAC	United States Air Force Environmental Technical Application Center
USCINCPAC	United States Commander in Chief, Pacific Command
USFJ	United States Forces, Japan
USN	United States Navy
USSPACECOM	United States Space Command

USSR	Union of Soviet Socialist Republics
VLO	very low observable ($\sim$ –25 to –30 db)
WAM	wide area munition
WMD	weapons of mass destruction

Bibliography

Aftergood, Steven. "A Revolution in Military Affairs?" F.A.S. Public Interest Report *Journal of the Federation o f American Scientists* 48, no. 1 (January/February 1995).

Air Force Manual (AFM) 1-1, *Basic Aerospace Doctrine of the United States Air Forc e.* 2 vols. Washington, D.C.: Department of the Air Force, March 1992.

Arquilla, John, and David Ronfeldt. "Cyberwar is Coming!" *Comparative Strategy* 12 (1993): 141–165.

Bacevich, A. J. "Preserving the Well-Bred Horse." *The National Interest.* Fall 1994.

Barnett, Jeffery R. "Nonstate War." *Marine Corps Gazette,* May 1994, 84–89.

Bodnar, John W. "The Military Technical Revolution: From Hardware to Information." *Naval War College Review* , Summer 1993.

Bowie, Christopher J., et al. *The New Calculus.* Santa Monica, Calif.: RAND, 1993.

____. *Trends in the Global Balance of Airpowe r*. Santa Monica, Calif.: RAND, 1995.

Bracken, Paul, and Raoul Henri Alcala. *Whither the RMA: Two Perspectives on Tomorrow's Army* . US Army War College: Strategic Studies Institute, 22 July 1994.

Bracken, Paul. "The Military After Next." *The Washington Quarterly,* Autumn 1993, 157.

Breemer, Jan S. "The End of Naval Strategy: Revolutionary Change and the Future of American Naval Power." *Strategic Review,* Spring 1994.

Caton, Jeffrey L. *Rapid Space Force Reconstitution* . Maxwell AFB, Ala.: Air University Press, 1994.

Clodfelter, Mark, and John M. Fawcett, Jr. "The RMA and Air Force Roles, Missions, and Doctrine." *Parameters,* Summer 1995.

Creveld, Martin van. *Technology and Wa r*. New York: Free Press, 1991.

Dick, Charles. "Russian Views on Future War—Part 2." *Jane's Intelligence Review*, October 1993, 451.

Dunn, Richard J., III. *From Gettysburg to the Gulf and Beyond : Coping with Revolutionary Technological Change in Land Warfare*. McNair Papers no. 13. Washington, D.C.: The Institute for National Strategic Studies, National Defense University, 1992.

Echevarria, Antulio J., and John M. Shaw. "The New Military Revolution: Post-Industrial Change." *Parameters*, Winter 1992–93.

FitzGerald, Mary C. *The Impact of the Military-Technical Revolution on Russian Military Affair s*. Washington, D.C.: Hudson Institute, 1993.

_____. "Russia's Vision of Air-Space War." *Air Force Magazine*, December 1993.

_____. "The Russian Image of Future War." *Comparative Strategy* 13, no. 2 (1994).

FitzSimonds, James R., and Jan M. van Tol. "Revolutions in Military Affairs." *Joint Force Quarterly*, Spring 1994.

FitzSimonds, James R. *The Revolution in Military Affairs : Challenges for Defense Intelligence*. Washington, D.C.: Consortium for the Study of Intelligence, 1995.

_____. "The Coming Military Revolution: Opportunities and Risks." *Parameters*, Summer 1995.

Garrity, Patrick J. *Why the Gulf War Still Matters: Foreig n Perspectives on the War and the Future of Internationa l Security*. Los Alamos, N. Mex.: Center for National Security Studies, July 1993.

Gray, Colin S. "Space Power Survivability." *Airpower Journal* 7, no. 4 (Winter 1993): 27.

Herman, Paul F. "The Military-Technical Revolution." *Defense Analysis* 10, no. 1 (April 1994): 91–95.

_____. "Future Military Powers." *Defense Analysis* 11, no. 1 (1995).

Hitchens, Theresa, and Robert Holzer. "DoD Eyes 21st Century Now." *Defense News*, March 28–3 April 1994, 1.

Holzer, Robert, and Stephen C. LeSueur. "DoD Plans Revolution in Technology, Tactics." *Defense News*, 23–29 May 1994, 1.
_____. "Pentagon Plans for Revolution." *Army Times*, 6 June 1994, 1.
_____. "A Revolution in War Tactics." *Navy Times*, 13 June 1994, 1.
Hundley, Richard O., and Eugene C. Gritton. *Future Technology-Driven Revolutions in Military Affair s*. Santa Monica, Calif.: RAND, 1994.

Jablonsky, David. "US Military Doctrine and the Revolution in Military Affairs." *Parameters*, Autumn 1994.
Joint Pub 3-56.1, *Command and Control for Joint Air Operations*. 14 November 1994.

Kendall, Frank. "Exploiting the Military Technical Revolution: A Concept for Joint Warfare." *Strategic Review*, Spring 1992.
Klaaren, Jonathan W., and Ronald S. Mitchell. "Nonlethal Technology and Airpower: A Winning Combination for Strategic Paralysis." *Airpower Journal* 9 (Special Edition, 1995): 42–51.
Krepinevich, Andrew F., Jr. "Keeping Pace with the Military-Technological Revolution." *Issues in Science and Technology* , Summer 1994.
_____. "Cavalry to Computer: The Patterns of Military Revolutions." *The National Interest*, Fall 1994.

MacGregor, Douglas A. "Future Battle: The Merging Levels of War." *Parameters*, Winter 1992–1993.
Mazarr, Michael J. *The Military Technical Revolutio n*. Washington, D.C.: Center for Strategic and International Studies, 1993.
_____. *The Revolution in Military Affairs: A Framework for Defens e Planning*. US Army War College: Strategic Studies Institute, 10 June 1994.
McKenzie, Kenneth F., Jr. "Beyond Luddites and Magicians: Examining the MTR." *Parameters*, Summer 1995.
Metz, Steven, and James Kievit. *The Revolution in Militar y Affairs and Conflict Short of Wa r*. US Army War College: Strategic Studies Institute, 25 July 1994.
Morocco, John D. "US Military Eyes Revolutionary Change." *Aviation Week & Space Technology*, 1 May 1995.

Morton, Oliver. "Defence Technology." *The Economist*, 10 June 1995, 5–20.

Ochmanek, David. "Military Technology: Revolution or Evolution?" Letter to the editor, *Issues in Science an d Technology*, Winter 1994–1995, 7.
Owens, William A. "JROC: Harnessing the Revolution in Military Affairs." *Joint Force Quarterly*, Summer 1995, 55–57.
_____. "The Emerging System of Systems." US Naval Institute *Proceedings*, May 1995, 35–39.

Pagonis, William G., with Jeffrey L. Cruikshank. *Moving Mountains*. Boston: Harvard Business School Press, 1992.
Perry, William J. "Desert Storm and Deterrence." *Foreign Affairs* 70 (Fall 1991): 66–82.
_____. Annual Report to the President and Congress, February 1995.
Pine, Art. "Pentagon Looks to Start High-Tech Revolution in Ways of War." *Los Angeles Times* (Washington edition), 27 July 1994.

Ricks, Thomas E. "Warning Shot: How Wars Are Fought Will Change Radically, Pentagon Planner Says." *Wall Street Journal*, 15 July 1994.

Senge, Peter M. *The Fifth Dimension*. New York: Doubleday, 1990.
Slipchenko, Vladimir I., Maj Gen, Retired. "A Russian Analysis of Warfare Leading to the Sixth Generation." *Field Artillery*, October 1993, 38.
Starr, Barbara. "Plotting Revolution at the Pentagon." *Jane's Defence Weekly*, 10 June 1995.
Stech, Frank J. "Sociopolitical Stresses and the RMA." *Parameters*, Summer 1995.
Sullivan, Gordon R., and James M. Dubik. *Land Warfare in the 21st Century*. US Army War College, February 1993.
_____. "War in the Information Age." *Military Review*, April 1994, 46–61.

Toffler, Alvin and Heidi. *War and Anti-War: Survival at the Daw n of the 21st Century*. Boston: Little, Brown, and Co., 1993.

Warden, John A., III. *The Air Campaign: Planning for Comba t.* Washington, D.C.: National Defense University Press, 1988.

_____. "The Enemy as a System." *Airpower Journal* 9, no. 1 (Spring 1995): 40–55.

Welch, Thomas. "Technology, Change and Security." *Washington Quarterly*, Spring 1990.

Index

Congo: 137–139, 146, 149
cruise missile: *xxiv*, 26–27, 31–32, 34–35, 37–41, 49, 52–53, 55, 58–59, 63–66, 73, 76, 78, 80, 86–87, 94–99, 101–104, 111, 117–123, 125–128, 130–132, 135, 138–141, 143–146, 148

D-ASAT: 45
Defense Intelligence Agency: 108
DIA: 62, 108

Eighth Air Force: 9, 44
electromagnetic pulse: 42, 79
EMP: 42–43, 45, 52, 55, 63, 79, 83–84, 86, 123, 125, 127, 130, 132, 136, 141, 145, 147

F-117: 34, 60
F-14: 93
F-15: 84
F-15E: 34
F-16: 99, 121, 140
F-18: 121, 140
F-22: 34, 92, 122, 141
Fogleman, Ronald R.: 20, 68
France: 14–15, 23, 67, 78, 135
frontline fighters: 93–94

GEO: 42, 45–46, 54, 66, 84, 103, 119, 123–126
Garcia, Diego: 109, 138, 140, 148
geosynchronous earth orbit: 42, 84
Germany: *xviii*, 14, 21, 67, 78
global positioning systems: *xix*, 27, 77
GPS: 27, 39–40, 43, 46, 66–67, 96, 99–100, 103, 117, 126–127, 135, 144–145, 147–148
Guam: 120, 122, 127, 130
Gulf War: *xi*, 4, 6, 9, 17, 60, 68–69, 91, 105, 109

Hanoi: 117
high-powered microwave: 42
HPM: 42, 125, 132, 136, 147
Hussein, Saddam: 74

ICBM: 52, 77, 125
information war: *ix, xix–xx,* 2–4, 8–9, 28, 73, 108–109
Iran: 73–74, 87
Iraq: *xix,* 9, 23, 71–74, 87, 91
IRBM: 135, 146

J-STARS: *xxvi,* 17, 48–49, 54, 59, 66, 68–69, 120, 125, 138,
    143–144
J3: 107, 109
J5: 107–110
J6: 109
Japan: *xviii,* 67, 77, 117–119, 122, 127
JFACC: *xxii,* 6, 36, 51, 54–55, 60–62, 65–66, 69, 96, 99,
    103–104, 116–119, 122–123, 126, 128–129, 131, 140–141,
    144–148
JFACC HQ: 141
JFICC: *xxii–xxiii,* 4, 49–50, 67, 85, 104, 122–123, 129, 141, 147
Joint Force Air Component Commander: *xxii,* 4, 49, 67, 85,
    116, 122, 129, 141

Kenya: 138, 140, 148
Khabarovsk: 115
Kinshasa: 135, 137–138, 140–141, 144, 146, 148
Komsomolsk: 115–116, 122
Korea: *xix,* 23, 28, 62, 68, 71–73, 77, 91, 139
Kuwait: 72

laser: 36, 38, 41–43, 91, 123, 125, 128, 129, 147
LEO: 43–46, 54, 66, 83–85, 103, 119, 124–126
logistic: 8, 19, 24–25, 38, 50, 53, 55, 57–58, 60, 67, 69,
    76–77, 81, 89–90, 129, 131, 140, 143, 148–149
low earth orbit, *xxvi,* 43, 136

magneto-hydrodynamic: 42
METT-T: 42
MHD: 42, 52, 86, 125, 127, 132, 136, 147
MiG-29: 93–135
Mitchell, William ("Billy"): 25
Mons: 83

National Command Authorities: *xxvi*
NCA: 45, 47, 52, 67, 119, 123, 131, 149
Nigeria: 73, 135–141, 143–149
Nitze, Paul: 11
nonlethal weapons: 45, 52, 55, 67, 69, 119, 127–129, 132, 142
Nonstate warfare: *xiii*
NSA: 62, 109
nuclear weapons: *xxiii*, 27–28, 52, 63, 77–78, 82, 102, 120, 149

OODA loop: 6–7, 39, 62–63
Operation Desert Storm: 9, 31, 57, 60, 64, 69, 102
Owens, Willam A.: 20

P-38s: 25
P-51s: 25
parallel war: *xxi*, 8–13, 110, 121, 131–132, 139, 149
Patriot: 30, 32, 38, 123, 138
Perry, William: 1
Petropavlovsk: 122, 125, 128, 130

revolution in military affairs: *xx–xxi*, 13, 15–16
RMA: *xx*, 13–15
Royal Air Force: 24
Russia: 2, 67, 78, 115, 120, 123, 127, 129–130, 136

SATCOM: 4, 46, 98, 122–123, 126, 141, 146
satellite: *xxii–xxiii*, *xxvi–xxvii*, 1, 4, 8, 40–52, 54–55, 61, 65–66, 69, 76, 78–80, 82–86, 99, 103–105, 111, 119, 121–126, 128–130, 135–136, 139–144, 147
Saudi Arabia: 68
Schwarzkopf, H. Norman: 83
Seoul: 117
Simulation: *xxii*, *xxvi*, 17–19, 28, 61
Slipchenko, Valdimir I.: 2, 20
Soviet Union: 21
space: *xxiii*, *xxv–xxvi*, 6, 12–13, 19, 27–28, 36, 41–43, 45, 47, 52, 54, 60, 62, 65–66, 68–69, 73, 80, 83–85, 91, 102, 111, 118, 121–126, 128, 139, 141–143, 146

stealth: *xxi–xxii, xxiv–xxv,* 12, 27–28, 31, 33–38, 41, 49, 52–53, 55, 57–61, 63–66, 81, 92, 96, 98–99, 102–104, 108, 110, 117, 121–122, 125–127, 130, 139, 141–142, 145–146
Su-24s: *xxiv*
Su-35: 135–136, 142
Sullivan, Gordon R.: 20

T-72: 144
Taiwan: 120, 130
tank: *xxv,* 11–12, 14, 16, 53, 59, 74, 77–78, 86–87, 89, 144, 148, 150
TLAM: 34
TR-1: 48–49, 125, 143
Treaty of Locarno (1925): 22
Trost, Carlisle A. H.: 68

U-2: 48
UAV: *xxiii, xxvi,* 8, 36, 39, 46–50, 53–55, 59, 66, 82–83, 85, 103, 111, 121–122, 124, 128–131, 140–142, 144–147
unmanned aerial vehicles: *xxiii,* 80
USSR: 23

Valdivostok: 115–117, 122
von Paulus: 100

Warden, John A.: *xv*
Whiteman AFB: 141
World War II: *xxiv,* 9, 14, 16, 22, 25–26, 38, 44–45, 64, 78
WWI: 64

Zaire: 135, 137–141, 143, 146, 148–150